Dynamic Soaring Dissected

How albatrosses and aircraft fly

Colin Taylor

Grosvenor House
Publishing Limited

The right of Colin Taylor to be identified as the author of this
work has been asserted in accordance with Section 78
of the Copyright, Designs and Patents Act 1988

The book cover is copyright to Colin Taylor

This book is published by
Grosvenor House Publishing Ltd
Link House
140 The Broadway, Tolworth, Surrey, KT6 7HT.
www.grosvenorhousepublishing.co.uk

A CIP record for this book
is available from the British Library

ISBN 978-1-80381-236-6

Contents

Chapter Heading Illustrations

Intro   Laysan albatross   CGI

        Slingsby T21 Sedbergh   Pencil sketch CT

1   Laysan albatross   CGI

2   Pelicans   CC

3   Schleicher K6e over the RAF College Cranwell   Oil on board CT

4   Digital model albatross   MB

5   LS7 sailplane hill-soaring at Camphill, Derbyshire   Oil on board CT

6   Royal albatross   CC

7   Laysan albatross   CGI

8   Laysan albatross   CGI

9   Laysan albatross   CGI

10   Laysan albatrosses   CGI

11   Puchacz (Eagle owl) glider at Derbyshire and Lancashire Gliding Club, Camphill   CT

12   Chipmunks   Watercolour CT

13   Laysan albatross   CGI

14   Model glider launch   Digital painting CT

15   Laysan albatross   CGI

Bird photographs from Creative Commons (CC)

Art works by Colin Taylor (CT)

Computer Graphic Images (CGI) are stills from 'Dynamic soaring The flight of the albatross' An animation on YouTube by Colin Taylor

Digital model albatross by CT with flow visualisation by Mikko Brunner (MB)

Introduction

The mechanism of flight is fascinating and well understood in relation to manned aviation but how do birds fly? Some aspects of bird flight, for example gliding, are similar to manned and model aviation but flapping is very different compared to propeller or jet flight. The speed and subtlety of bird flapping and manoeuvring is very difficult to read.

In recent years and particularly since the TV series Life of Birds in 1998 and later Earthflight were broadcast, recording technology has progressed to the point that the motion and movement of birds in free flight can be analysed in exquisite detail, in real time and in slow motion. This started me on an exploration of flapping flight, particularly the complex motion of the articulated bird wing. I have built several successful model ornithopters (flapping wing aircraft) but, to be fair, they really work more like big bugs than small birds. I have also built working models of the articulated bird wing in order to understand its complex folding geometry but there is considerable difficulty in delivering controlled power to what is, of necessity, a light-weight and fragile structure. Nevertheless, this study has produced some insights concerning the way that birds use the articulated wing geometry for thrust, control and stowage.

In parallel with this, a study of the literature revealed very little detail about how birds manage power and control; theory of flight forms a very small part of ornithology. Explanations of how thermal soaring and hill soaring are accomplished by birds are satisfactorily dealt with, being the same as manned soaring. However, dynamic soaring, the flight style of the albatrosses and giant petrels, is quite unlike anything in manned aviation. I found dynamic soaring poorly explained and the diagrams supporting the explanations appeared to be quite improbable and unrealistic. On the face of it, the traditional explanation for dynamic soaring, the wind gradient theory, is quite plausible and easy to understand but it did not fit well with my understanding of how to fly an aircraft and how aircraft behave in turbulence and wind-shear.

Then, since the 1990's, radio-control model glider pilots began flying their machines in the lee of hills in circling flight, maintaining average position and height but gaining huge speeds exceeding 500mph. This is also described as dynamic soaring but clearly it is not exactly the same as what the albatrosses are doing. The aeromodellers were doing this intuitively and again the wind-gradient is cited as the mechanism enabling this activity. But once more, the theory does not really explain what is going on.

There is more to dynamic soaring than simply climbing and descending in a wind gradient. In particular there needs to be a reason why speed increases in a downwind turn and reduces in an upwind turn. There have to be accelerations and forces involved. Also, there has to be a mechanism to exchange momentum and kinetic energy between the glider and the wind and both the birds and the model gliders must be governed by the same rules of motion and aerodynamics. New filming techniques and high resolution satellite tracking of albatrosses have now revealed the true shape of their dynamic soaring manoeuvres and give a clearer view of what needs to be explained. Dynamic soaring is a complex subject entangling the threads of physics, mathematics, aeronautics and biology. It is not only about how birds and gliders fly but also about how everything flies because all flying is governed by the same rules of dynamics.

When I had a partial theory of dynamic soaring, I published my results in a website called dynamic-soaring-for-birds, to try and find other people interested in the subject and to provoke a debate. This did not go down well! However, there was some correspondence and the questions posed and general incomprehension displayed did provoke me to find a solution. Eventually, I was able to explain both the upwind and downwind turns and the exchange of momentum and I have developed a mathematical model which explains the flight of both the albatrosses and the model gliders. I then produced a computer animation entitled 'Dynamic soaring. The flight of the albatross' which is on YouTube. This illustrates the dynamic soaring mechanism graphically. Now, this book is a collation of everything I have learned about dynamic soaring. It is a pilot's view of flying in the natural world and in the world of model aircraft and full-size flying machines.

In re-writing the story of dynamic soaring, I will be ruffling the feathers of a few pilots and others who may struggle to believe that there is something new to learn about flying. However, I am bringing to the game over 50 years of flying experience. I learned to fly when I was at school and in the Air Training Corps and I have been flying and gliding in one capacity or another ever since, as private pilot, flying instructor, commercial pilot and glider pilot. For me, knowing how to fly was an important part of analysing and understanding dynamic soaring, because it is all about the way the manoeuvre is flown and the practical effects of handling an aircraft in a wind. All the time I have been watching the sky and watching the birds flying there. I am asking: what do we have in common and what are the differences between bird-flight and manned aviation? To understand how dynamic soaring works it is necessary to take apart the manoeuvre and examine it with a pilot's eye; then reassemble it with some mathematical glue to make sure it all sticks together.

Colin Taylor

Timperley

2022

Chapter 1

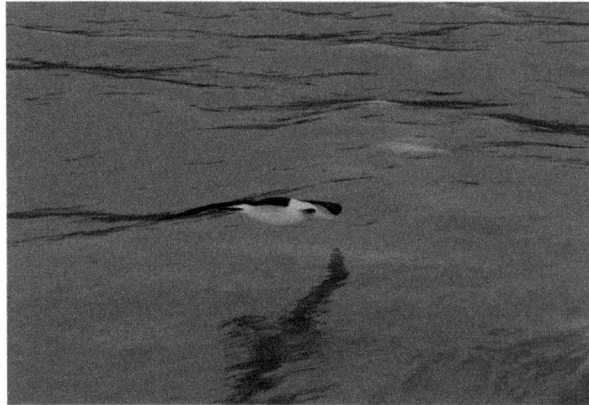

Riding the Wind

The air is so insubstantial – hardly able to support a falling feather or a fleck of sea-foam. Yet the wind is literally a force of nature able to lift a sea-swell or knock-down a building. The wind can sap the energy of a careless flyer; yet albatrosses are able to harness the wind and turn it to their advantage. Through the stadium of the Earth's boundary layer soars the marathon bird, sabre-wings slashing the breeze, feather-tips stroking the water. Now skimming the surface, hardly bothering the sky; next, zooming, tilting on a wing-tip in sinuous, undulating flight. Ever watchful, pressure and scent sensitive; seeking the aerodynamic sweet-spot and the pungent air-path, the crumb-trail to the next meal. With a flick of the wing in rhythmic sequence they ride the wind for thousands of miles and even for a lifetime - the oldest recorded Laysan albatross is thought to be 70 years old.

The flight style of the albatross is known as dynamic soaring. However, dynamic soaring means different things to different people. To an ornithologist it is the way that albatrosses fly, how they use a winding flight pattern to maintain average speed and height over distances of hundreds of miles. To a radio-control pilot, it is the way that a model-glider can be made to fly at huge speeds, hundreds of miles per hour, maintaining height in circling flight on the lee-side of hills. In both cases, dynamic soaring uses the horizontal motion of the air to sustain height and speed in gliding flight. It is an active process in which the actual manoeuvre, the way the aircraft or bird turns, facilitates an exchange of momentum and energy between the wind and the flying machine. On the other hand, conventional soaring uses the vertical motion of the air to sustain the aircraft in flight; a passive process in which the glider flies in equilibrium or in a steady turn and relies on the air going up faster than the glider is going down.

Then again, there are all the other flying creatures and flying machines which are not specifically trying to dynamic soar and which consequently do not gain any advantage but instead haplessly lose energy because of the effect of the wind whilst turning. This third group may not even be aware that they are losing energy and may take some convincing of this.

The Wind Gradient Theory

If you are familiar with this topic you may ask: Do we not have an explanation for dynamic soaring dating from the 19th century? Despite the different flight manoeuvres performed by the birds and the model-glider pilots, both will typically be explained by the same method: The wind-gradient theory or the Rayleigh Cycle, first expounded by Lord Rayleigh in 1883 in the journal Nature. Virtually all works on dynamic soaring refer to Lord Rayleigh; his explanation for the soaring of birds has been analysed mathematically and developed in various ways, so that it seems like a done-deal.

Unfortunately, the wind-gradient theory on its own does not fully explain what is going on in dynamic soaring. It contains certain assumptions about the shape of the manoeuvre and the properties of the wind and does not take into account what happens when the flying machine or creature is turning in a wind. Chapter 2 contains a critique of Lord Rayleigh's article and, to some extent, a de-bunking of the wind-gradient theory of dynamic soaring.

Turning in a wind

This all leads us to a question: Is there something which needs to be explained, which is more fundamental than the beneficial soaring exhibited by the albatrosses and the model gliders? The answer is yes, there is a basic effect which is to do with aircraft turning in a wind; an effect which can be turned to advantage by birds and aeromodellers performing particular manoeuvres but which has, on average, a negative effect when a normal turn is made. To explain dynamic soaring, we will analyse what happens when an aircraft turns in a wind; something which falls between the two disciplines of aerodynamic theory and flying technique and which is normally omitted from both.

Can it really be true, that the effect of the wind on an aircraft while turning has been overlooked ever since the Wright brothers learned to fly their glider in 1902 or indeed since George Cayley flew his first glider in 1804? It would appear to be so, although to be fair, the basic effect is completely automatic and requires no input from the pilot and so is easily missed. The basic effect of the wind occurs every time a flying creature, whether natural or mechanical, turns in a wind because all flying machines and flying creatures are subject to the same rules of dynamics and aerodynamics. It normally results in a small loss of energy, either a loss of airspeed or height; a loss so small that it is easily overlooked or hidden among the other losses incurred while turning. Only if some benefit is to be gained from the wind while turning does the bird or the pilot need to do some specific manoeuvre. It appears that both the albatrosses and the model-glider pilots developed their dynamic soaring techniques by instinct and intuition alone, without a comprehensive theory to guide them.

Theory of flight

Conventional theory of flight covers circumstances when the aircraft is in equilibrium and is flying in still-air; that is not under acceleration except when turning. The forces acting on an aircraft are explained by relative airflow, while assuming straight and level flight or steady rates of climb and descent or turning flight in still air, in which ground-velocity is the same as air-velocity and with steady forces giving a uniform rate of turn. In other words, the effect of the wind is disregarded and considered to be irrelevant so far as the aerodynamic forces are concerned. And to be clear, when I refer to the wind I am talking specifically about the speed of the air relative to the ground. Airspeed is the relative speed of the aircraft and the air.

The effect of the wind is also disregarded during the initial lessons of conventional flying training, which are to do with the effects of controls and the basic manoeuvres of level flight, climbing, descending and turning. The wind becomes relevant when the pilot starts to learn to take-off and land because headwinds and tailwinds affect groundspeeds and glide-angles, while drift-angles affect the take-off and landing directional control techniques. However, conventional theory of flight or theory of navigation does not explain exactly how and why groundspeed changes when an aircraft turns relative to the wind.

Chapters 3 to 7 will explain the theory of flight in aircraft and birds. For completeness, we will look at gliding, soaring and flapping flight and we will look at navigation because the triangle-of-velocities is an important part of understanding how dynamic soaring works. This will be based on my experience of flight and of building and flying model aircraft including ornithopters.

Animal behaviour

Dynamic soaring is not only about aircraft dynamics but also about animal behaviour - what albatrosses do instinctively and what model-glider pilots do intuitively. Most people, from their day-to-day observations, have some idea of what aircraft and birds do. That is a good starting point for explaining how aircraft and birds fly. However, few people have seen albatross dynamic soaring and even fewer know the true nature and shape of the manoeuvres these birds perform. This is because albatrosses fly over distant oceans and are therefore rarely seen in pure dynamic soaring flight. Also, when they are seen, it is normally from the deck of a ship with a limited view of the bird's flight path which is a three-dimensional manoeuvre covering several hundred metres. Before attempting to explain dynamic soaring, we need to ask: What is it that we are trying to explain? What is it that albatrosses actually do? The answer to this will be found in Chapters 8 and 9 which analyse data from GPS tracking of the birds. This is essential if we are to understand how albatrosses do dynamic soaring; it is not enough to just assume or guess what the shape of a dynamic soaring manoeuvre is like. Chapter 10 contains a narrative description of the dynamic soaring manoeuvre.

The Basic Effect of the wind

What is this basic effect of the wind on airspeed? Simply this: the airspeed of the aircraft will be affected by any *rate of change* of the headwind component, while inertia will resist any change to the actual-speed or groundspeed of the aircraft. This is made manifest when climbing or descending through wind-gradients, or whenever the aircraft is affected by turbulence *or while turning relative to the wind direction*. Additionally, the basic effect of the wind is to cause an angle of drift which allows aerodynamic force components to accelerate the bird or aircraft in two different directions; the directions of the air-velocity and the ground-velocity.

The effect of the wind-gradient has been readily accepted by theorists and students of dynamic soaring, while the effect of wind-shear and turbulence is well known to all pilots. However, the effect of the wind during accelerated and turning flight is less easy to understand than the other two effects and this is what has been overlooked. How do the aerodynamic forces and the effect of the wind generate an excess of speed and height at the end of each dynamic soaring manoeuvre?

Explaining all of this will be the main objective in chapters 11, 12 and 13. The results of calculations which model these effects are shown in graphical form, including a model of the albatross dynamic soaring manoeuvre.

The Windward Turn Theory

I call the explanation of dynamic soaring the Windward Turn Theory. It will introduce a novel way of looking at the forces and accelerations acting on an aircraft in flight and will inevitably involve both traditional theory of flight and aspects of practical flying technique. I am not saying that these ideas are completely new but I think this is the first time the various threads have been brought together in one coherent theory.

The Windward Turn Theory explains why groundspeed changes when turning in a wind and how the airspeed is also affected. It explains why albatrosses fly the way they do; how dynamic soaring defines their physiology and how it fits-in with their foraging strategy; how albatrosses are able to dynamic soar upwind, downwind and crosswind. Also, it explains how radio-control gliders are able to achieve high speeds in circling flight on the downwind side of hills. It explains how momentum and energy are exchanged between the turning aircraft and the wind. The theory explains the true role of the wind gradient but does not always depend upon it. It will also explain in general terms how the wind affects an aircraft while turning, why there is normally an overall loss of energy when turning in a wind and it will partly explain the long-standing myth of the downwind-turn and its role in aircraft accidents.

RC-glider dynamic soaring is discussed in chapter 14. Chapter 15 describes albatross dynamic soaring upwind and downwind, which is closely related to what the model gliders do. Finally, Appendix 1 contains a description of the mathematical model that produced the graphs.

So, how did it all begin; what did Lord Rayleigh actually say about dynamic soaring?

Chapter 2

The Soaring of Birds

Lord Rayleigh and the Wind Gradient Theory

Wherever you look for references to dynamic soaring, whether on the internet or in Wikipedia or in books on ornithology or on the mechanics of bird flight, you will find reference to Lord Rayleigh's article **THE SOARING OF BIRDS** in **Nature** of **April 1883**.

Lord Rayleigh was John Strutt, 1842-1919, a well-regarded scientist and educator. He was interested in gases and energy and won the Nobel prize for the discovery of argon gas. He presumably knew something of the atmosphere, the weather and the experience of balloonists. He writes with such authority and confidence that some authors say that ever since that time dynamic soaring has been understood and variations of his theory are repeated almost without question.

In 1883 the industrial revolution was well under way and the scientific method was well established but the study of flight was in its infancy. There had been precious little observation or experimentation on flight and, of course, nobody knew how to fly. Before 1883, the only person to have 'flown' a heavier-than-air craft was George Cayley's somewhat reluctant coachman in 1804. The earliest pioneers of manned flight were yet to take to the air; Otto Lilienthal began gliding in 1894 and the Wright brothers achieved their first glides in 1902.

The article was a response to a letter from Mr S E Peel who had observed, in Assam, pelicans and other large birds soaring in circles to heights of 8000 ft. Peel included several illustrations showing how he braced his gun in the forked branch of a tree and sighted on the birds to observe how they gained height as they circled and drifted downwind. Rayleigh's article is not long or detailed and does not attempt to explain the flight of the albatross in particular; nor is the term dynamic soaring used. Rather, it is an early attempt to explain avian soaring in general. His theory describes a kind of dynamic soaring in which the birds exploit a wind shear or wind gradient while circling.

So, what did Lord Rayleigh actually write in his 1883 article in Nature? He opens with this,

'I premise that if we know anything about mechanics it is certain that a bird without working his wings cannot, either in still air or in a uniform horizontal wind, maintain his level indefinitely.... Whenever ...a bird pursues his course for some time without working his wings, we must conclude either (1) that the course is not horizontal, (2) that the wind is not horizontal, or (3) that the wind is not uniform.

This statement recognises that (1) gliding involves a loss of height and that (2) a wind with a vertical component could cause a bird to rise. Clause (3) is strictly correct but requires a little more thought. If the wind is uniform then there is no change of wind-speed, momentum or energy. Therefore, there is no exchange of momentum between the wind and the bird and no energy is available to the bird to overcome drag or allow the bird to do what it wants to do. However, if the bird is able to exchange horizontal momentum with the wind, then the wind (or at least small packets of it) will not be uniform. In fact, this is exactly what does happen when an aircraft turns in a wind but in 1883 this was beyond their knowledge.

Referring to Peel's observations of pelicans soaring in circles, Rayleigh says this may happen when there is a wind and,

'That birds do not soar when there is no wind is what we might suppose, but it is not evident how the existence of a wind helps the matter. If the wind were horizontal and uniform it certainly could not do so'.

He continues,

'As it does not seem probable that at a moderate distance from the ground there could be a sufficient vertical motion of the air to sustain the birds, we are led to inquire whether anything can be made of horizontal velocities which we know to exist at different levels'.

This was a surprising thing to say. His earlier statement (2) seems to recognise the possibility of non-horizontal winds and a careful observation of a developing cumulus cloud *'at a moderate distance from the ground'* would have revealed the rising air-currents. A casual glance at a bird gliding straight, at constant height, on the upwind side of a hill or a stand of trees, reveals the fact that it must be flying in rising air. The horizontal velocities of the wind at different levels were probably the experience of the balloonists of the day. They found that as they drifted downwind and gained height, their track over the ground turned to the right (in the northern hemisphere) and as they descended the track turned to the left. This is due to the natural variation of wind-direction and speed with height.

We now know that what Peel observed was thermal soaring by birds circling in rising columns or bubbles of air, known today as thermals. If the air is rising faster than the bird is descending in the air, then the bird will gain height. The vertical currents of air which the birds are exploiting are caused by surface heating and atmospheric instability and are triggered by surface effects like localised solar heating or orographic features. Thermals do not require a wind, although they will normally generate a local wind near the surface due to the inflow of surrounding air to replace the air which is rising. Thermals not only enable birds to soar but also lead to the formation of cumulus clouds, close observation of which will reveal the atmospheric motion. It does seem extraordinary that the gentlemen involved in this correspondence were so blind to the possibility of strong vertical motion of air in the atmosphere being an explanation for the soaring of birds.

Rayleigh then describes what we nowadays call a kind of dynamic soaring. He says that,

'In a uniform wind the available energy at the disposal of the bird depends upon his velocity relative to the air about him'.

He is assuming that the kinetic energy of the bird is proportional to airspeed so that, if it has excess airspeed, it can zoom into a climb and gain height just like on a roller-coaster. This is true up to a point. In still air, airspeed and actual speed are the same, but in practice the airspeed gives the kinetic energy of the relative airflow from which are derived values of lift and drag. Also, increased velocity relative to the air means increased drag, requiring a steeper angle of descent to overcome the drag.

The kinetic energy of an aircraft is arguably proportional to its actual speed (squared) which is revealed when it contacts the ground and the energy must be dissipated in brake heat energy or something more destructive! This is a debatable point because speed and kinetic energy depend upon the frame of reference against which the speed is measured. In reality, the energy available to a glider is height. When the glider loses height at constant airspeed and rate of descent, the drag losses are directly equivalent to the height lost.

Having dismissed the possibility of soaring in a uniform wind, Rayleigh then proposes a non-uniform wind by describing the wind as a two-layer system, with the two layers moving at different speeds. But does that constitute a non-uniform wind in the context of dynamic soaring? The theoretical two-layer system is a consequence of the interaction of the wind and the ground. The wind loses speed energy because the lowest part of the atmosphere slows-down due to friction between the air and the ground, the energy being dissipated in the form of air turbulence, wind noise, the swaying of trees, the rising of waves and so on. All of this happens before the bird gets involved.

The Rayleigh cycle involves the bird gliding downwind and descending through a horizontal shear-boundary or plane, where the wind-speed reduces. He does not describe an exchange of momentum between the wind and the bird; the two wind-layers and the bird are all moving with constant velocity relative to each other. He goes on to say that,

'In falling down to the level of the plane there is a gain of relative velocity, but this is of no significance for the present purpose, as it is purchased by the loss of elevation'

Not necessarily. Because of the effect of drag, when an aircraft loses height it does not necessarily gain speed. In a descent, airspeed is maintained because aerodynamic drag is balanced by a component of weight. Drag energy losses, at constant airspeed and rate of descent, are equivalent to loss of height energy. Once drag is overcome at a particular angle of descent, a further increase in dive-angle will result in an increase in airspeed until a new balance of weight and drag is achieved.

'...but in passing through the plane there is a really effective gain. In entering the lower stratum the actual velocity is indeed unaltered, but the velocity relatively to the surrounding air is increased. The bird must now wheel round in the lower stratum until the direction of motion is to windward, and then return to the upper stratum, in entering which there is a second increment of relative velocity. ....if the successive increments of relative velocity squared are large enough to outweigh the inevitable waste which is in progress all the while, the bird may maintain his level, and even increase his available energy, without doing a stroke of work'.

This cycle is then repeated by circling to explain Peel's observation. Rayleigh says that actual velocity is unaltered meaning that inertia (mass) resists any change to actual speed but this must mean that momentum and kinetic energy of the bird depend on actual speed and not airspeed.

The problem with this is that, when descending through the shear layer, if actual speed (groundspeed) is preserved and the speed of the wind in each layer is constant, then there is no acceleration; kinetic energy derived from actual speed is unchanged. There is no gain of energy, only an increase in airspeed and drag and a loss of height. The only way to sustain the increased airspeed and consequently greater drag is to dive more steeply and use up potential energy more quickly. Wheeling around will result in further loss of actual speed or height.

The *'second increment of relative velocity'* during the upwind climb, will again increase drag and reduce actual speed and therefore reduce kinetic energy. It can only be achieved by firstly converting actual speed to height by using momentum to do work against gravity; but that means that if actual speed reduces, then the gain of airspeed must be less than the change of wind-speed. Note that Rayleigh avoids referring to constant actual speed when gaining height when returning to the upper stratum, because this would be impossible due to conservation of energy.

Does airspeed increase in the way that Rayleigh describes? When descending downwind through a wind-shear, airspeed may well increase suddenly but the increased drag-load will cause the actual-speed to rapidly reduce at the same time. Therefore, actual speed will not be constant and airspeed will not increase by the amount that the wind-speed changes. The unbalanced drag-load will then cause the airspeed to reduce to the original point of equilibrium. Actual velocity will only be constant if the angle and rate of descent is increased, allowing gravity to overcome the extra drag with corresponding loss of height.

When climbing upwind through a wind shear, actual velocity cannot be constant because work is being done against gravity. To gain height, there has to be a gain of potential energy and therefore a loss of kinetic energy and therefore a loss of actual speed. Again, the airspeed may increase suddenly due to turbulence or wind-shear but the increase in drag must cause the actual speed to reduce more quickly. Airspeed will not necessarily increase but, with gain of height and an increase of headwind, the loss of airspeed may be less than in still air.

Later in the article, referring to the difference of wind-speed with height, he recognises that,

'there is of course no such abrupt transition in nature ...there is usually a continuous increase of velocity with height...

which is true. He then says,

'...it is only necessary for him [the bird] to descend while moving to leeward and to ascend while moving to windward...'

But this does not explain Peel's observations of circling pelicans;

'Mr Peel makes no mention of the circular sweeps being inclined to the horizon...

Indeed, he does not. We now know that birds or gliders in a thermal, as were Peel's pelicans, gain height continuously and not by climbing and descending.

Later he writes,

'A priori I should not have supposed that the variation of [wind] velocity with height to be adequate for the purpose; but if the facts are correct [Peel's observation], some explanation is badly wanted'.

Rayleigh is saying that he thinks that the wind gradient is insufficient to enable the flight of the pelican and therefore the wind gradient theory itself is impracticable.

After a few more letters from Mr Froude, Mr Airy and Mr Baines, the debate had extended to include the flight of albatrosses. In 1898 Rayleigh again wrote to Nature and made the connection between his wind gradient theory and the soaring flight of albatrosses. Rayleigh's soaring model appears to relate to albatross flight because of the up and down motion of the birds in the supposed wind gradient, close to the ocean surface. The sailors of the time, and of now, are well aware of the wind gradients above the sea which create wind shadows in the troughs between swells.

However, albatrosses in dynamic soaring do not circle or make 180 degree turns as described in the Rayleigh cycle and do not necessarily get anywhere near to an upwind or downwind heading; although they are able to dynamic soar upwind. Observers of the time did not see or understand any of this.

Clearly, Rayleigh knew there was a weakness in his argument but he was not offering a definitive explanation of avian soaring. Rather he was just contributing to a debate and inviting the world to provide the answer. In a sense he was correct, that a bird cannot soar without a changing wind. However, as explained later by the Windward Turn Theory, a uniform horizontal wind is simply a wind with a particular velocity at the time and place it is encountered by the bird. The non-uniformity of the wind required by Rayleigh's statement (3), is not an intrinsic part of the wind itself, a wind gradient, but rather an acceleration of the headwind component experienced by the bird as a consequence of the way the bird turns relative to the wind. The variation of the wind is then caused by the exchange of momentum, back and forth, between the wind and the bird as the bird turns and encounters each successive unit mass of air. In chapter 10, The Windward Turn Theory explains how the aerodynamic forces affect the acceleration of the bird; which is seen as a rate of change of groundspeed and thus a rate of change of momentum and kinetic energy exchanged with the wind.

Since 1883 and principally since the end of the First World War, many different kinds of soaring have been flown and described including thermal soaring, hill soaring and atmospheric wave soaring. With the development of high-altitude flight and the discovery of atmospheric waves in the lee of mountains, the toy-box of soaring techniques was complete. The Rayleigh cycle has been left to explain dynamic soaring as practised by albatrosses. The pity is that that the world has taken his contribution to the debate to be the whole answer and has not really completed the dialogue, until now.

What is wrong with the Wind-gradient theory?

There is no doubt that when an aircraft encounters turbulence or wind gradients they do have an effect, particularly on airspeed while inertia resists any change to actual speed. The Wind-gradient theory is plausible and is supposed to work like this: The glider descends downwind and passes through a horizontal shear boundary into a layer of slower or stationary air. Actual speed is constant and airspeed increases by the same amount as the change of wind-speed. The glider then turns onto an upwind heading, climbs back up through the shear boundary and the airspeed again increases by the same amount as the change of the wind-speed. The glider then turns downwind and repeats the process. If the gain of airspeed in the wind gradient is equal to or greater than the losses due to drag in the turns, then the excess airspeed converts to height and height can be maintained or gained.

2.1 Rayleigh Cycle

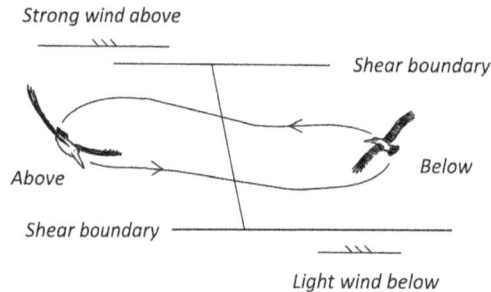

IF this is so then, given a representative airspeed, lift/drag ratio and angle of bank, the airspeed lost in the turns can be calculated. The airspeed lost then equals the wind gradient and, assuming a two-layer wind system, the wind is calculated.

But this is just a self-fulfilling proposition following the same faulty logic which is that airspeed can be gained in the wind-gradient which equals airspeed lost in the turns. It does not describe a practical dynamic soaring mechanism involving a transfer of momentum and energy between the wind and the bird. In the wind-gradient theory, the bird gains airspeed energy but the wind does not lose energy. And dynamic soaring is not energy-neutral as some people assert, because there is a net loss of energy on the part of the bird in the form of aerodynamic drag. The wind gradient does indeed represent a loss of wind energy but that happened before the bird arrived and not as an exchange of energy with the bird.

The wind gradient theory is not practical

If a bird passes from one layer of air to another, maintaining constant actual-velocity, and each layer is moving at constant velocity, then the speed of the bird relative to each layer, remains the same. There is no acceleration of the mass of the bird and no gain of energy based on actual speed. The only change is a different value of airspeed and drag in each layer.

When descending downwind through a wind-shear, airspeed may well increase suddenly but the increased drag-load will cause the actual-speed to reduce at the same time. Therefore, actual speed will not be constant and airspeed will not increase by the amount that the wind-speed changes. The unbalanced drag-load will then cause the airspeed to reduce to the original point of equilibrium. Actual velocity will only be constant if the angle and rate of descent is increased, allowing gravity to overcome the extra drag with corresponding loss of height.

When climbing upwind through a wind shear, actual velocity cannot be constant because work is being done against gravity. To gain height, there has to be a gain of potential energy and therefore a loss of kinetic energy and therefore a loss of actual speed. Furthermore, if the change of height is the same in climbing and descending through the shear boundary, but the actual speed is greater in the downwind descent, then the same amount of PE and KE is only worth a smaller speed increment at the greater downwind speed, because kinetic energy is a square law. Therefore, overall, actual speed or height is lost.

The theory is not a good model of the atmosphere

In Rayleigh's model of soaring pelicans, to achieve height gain the thin shear boundary would have to be repeated continuously as the bird climbs, which is not a good model of actual atmospheric wind gradients. Actual wind gradients are quite modest and marked shear boundaries normally only occur near to solid objects. Lord Rayleigh knew this and said as much in his 1883 paper;

'A priori I should not have supposed that the variation of velocity with height to be adequate for the purpose;'

The wind gradient theory as described by Rayleigh and developed by others, assumes that between moving air layers there are improbably thin shear boundaries and therefore improbably steep wind gradients. It assumes the bird can climb and descend through a shear boundary with no gain or loss of height, no exchange of PE and KE, with no turbulence and no drag losses. Although airspeed and drag may increase there is no increase in total kinetic energy and no extra force in the direction of flight to overcome the extra drag.

In such a system, the stationary lower layer of the wind would need to be deep enough to accommodate a large bird in a banked turn, which could only occur occasionally in the lee of a breaking wave and not on a continuous basis. Any such calms in the troughs between swells will be moving with the swells and most probably downwind and so would not be still air.

The theory does not explain how actual speed increases in the downwind turn

Actual speed increases when the downwind turn is made in the upper layer and this is required by the Rayleigh cycle but is just taken for granted. The increase in actual speed requires a force which is not explained. If the force is gravity, then height will be lost which would represent a loss of energy.

It turns out that, with an angle of bank, a component of aerodynamic lift combined with the angle of drift, is responsible for the acceleration of groundspeed. The same mechanism is responsible for the loss of groundspeed during the upwind turn as explained later in the Windward Turn Theory.

GPS tracking of albatrosses (see chapter 8) shows periods when both the ground-speed and height are increasing at the same time. The wind-gradient theory does not explain this.

The theory is not a good model of bird flight

The theory does not explain what the pelicans are doing in thermal soaring. They are not losing height to gain height. They are gaining height continuously in rising air.

Nor is it a good model of albatross dynamic soaring in which the windward and leeward turns are of a distinctly different shape. The Rayleigh cycle assumes either 360 degree turns or alternating 180 degree turns to left and right but albatrosses do not do that. In albatross dynamic soaring the amount of turn is approximately crosswind plus and minus 20 to 30 degrees and is nowhere near 180 degrees. Any angle off the wind reduces the headwind or tailwind component and reduces the wind-gradient by the same amount.

For example, at 60 degrees off the wind, the headwind component, proportional to the cosine of the wind-angle, is half the actual wind and the effective wind-gradient is only half the actual wind-gradient.

It does not reflect the practicalities of actual flight

Few pilots have experience of descending downwind through wind-shears because we normally take-off and land into wind. My experience in light aircraft, of taking-off and climbing in strong headwinds, where the headwind increases by 10 to 20kts during a climb to 1000ft, is that the normal pitch attitude yields a slightly greater airspeed than in still-air; around 72 kts instead of 70 kts. This is sustained during the first 500ft of climb until the aircraft is turned away from the wind. In other words, during this climb, the airspeed does not increase by the amount the wind changes. There is an incremental increase of airspeed causing an unbalanced drag load and consequently a continuous reduction of ground-speed. I do not see either constant ground-speed or continuously increasing airspeed as I climb up through a wind gradient.

When landing into wind, if the aircraft is over-rotated in the flare with power at idle and starts to gain height through the wind gradient, the airspeed does not increase. On the contrary, as speed converts to height, the airspeed rapidly reduces as the inertia of the aircraft does work against gravity. The loss of airspeed then results in a rapid loss of height and a hard landing, if power is not increased to rescue the situation. Every student pilot and flying instructor has seen this situation!

Why is it thought that dynamic soaring is about the use of the wind gradient?

Perhaps this misunderstanding about the role of the wind gradient has come about because the observer's view of dynamic soaring is normally from the deck of a ship where the vertical motion of the bird is easy to see but the left and right turns are less obvious.

The wind-gradient can improve the efficiency of the albatross leeward turn flown as a wing-over, by reducing airspeed losses but without the wind-gradient being the actual source of energy. The ground-effect can improve the efficiency of the windward turn and, combined with the wind gradient and the bird's desire to achieve distance rather than height, is the reason the bird stays at low-level.

The fact is that the wind gradient is a reality, at least some of the time, and albatrosses do fly up and down through it, so you may prefer to leave it at that. Ignorance is bliss, as they say, and nobody is going flying over the ocean like an albatross any time soon - unless they are prepared to get their feet wet! Nevertheless, an understanding of how dynamic soaring works will lead to a deeper understanding of the effect of the wind on flight but it will be more complicated, so hang on to your hat!

Having disposed of the 19th century misconceptions about flight, in the next chapter we will look at what we definitely know about how aircraft fly.

Chapter 3

Theory of Flight

Introduction

This chapter will cover some of the principles of aerodynamics and theory of flight. Just enough to understand the dynamic soaring theory. To be a pilot you need to know about these subjects but not to the extent needed to research aeronautics or design aircraft. To be a birder you don't need to know about any of this but I hope this little bit of knowledge will help you better appreciate what birds do.

When I learned to fly in the early 1970's, Bernoulli's theorem was the preferred explanation for aerodynamic effects but it turns out there is more to it than that. There are several ways of explaining how aerodynamic forces are produced. Bernoulli explains the pressure difference around a wing but does not help to explain dynamic soaring. The Newtonian explanation of aerodynamic force is a more useful tool and leads neatly to the exchange of momentum and energy between the flying machine and the wind, which is essential to understand dynamic soaring.

Angle of attack

Angle of attack is the angle between the relative airflow and the chord line of the wing. The chord line is the line joining the leading edge and trailing edge. The direction of the relative airflow is its direction at a distance from the wing before the air is disturbed by the wing. When the wing section is cambered, at zero angle of attack there is still a positive lift coefficient. For cambered wings, zero lift occurs at a negative angle of attack. When the wing has a symmetrical section, as for example, on aerobatic aircraft, which need to fly upside down as well as upright, the lift coefficient will be zero at zero angle of attack.

Straight and level flight

There are four fundamental forces affecting an aircraft in flight; lift, weight, thrust and drag. In straight and level flight the aircraft is in equilibrium, that is to say not subject to acceleration. Lift acts normal to the direction of flight. In level flight, lift is equal and opposite to weight; with lift acting vertically upward and weight acting vertically downward. Thrust is equal and opposite to drag with drag being opposite to the direction of flight. (Figure 3.1).

Any unbalanced forces will result in acceleration in the direction of the force; so that excess thrust, for example, will cause the aircraft to increase speed. As speed increases, drag will increase until drag equals thrust and a new state of equilibrium is achieved. As airspeed increases, lift is maintained equal to weight by reducing the angle of attack. At high speeds a smaller angle of attack and a lower nose attitude is required.

3.1 Equilibrium

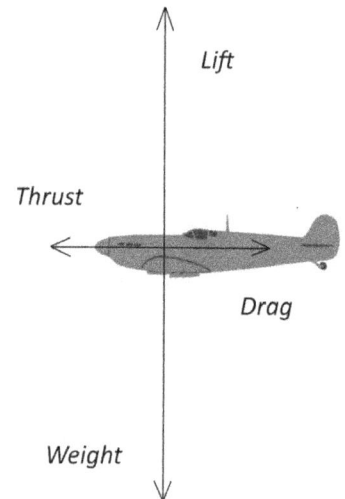

Gliding

3.2 Glider Equilibrium

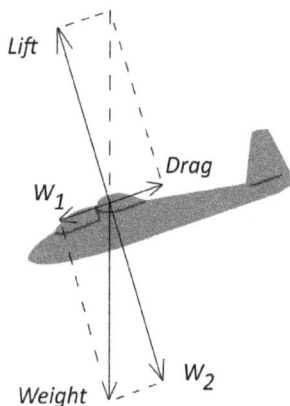

A glider has no thrust but the aircraft is essentially pointing downward and propelled by gravity just like a wagon rolling downhill – the steeper the hill the faster the wagon goes but the greater the drag. Drag and lift are still respectively opposite to and normal to the relative airflow but now they are balanced by components of weight W_1 and W_2. (Figure 3.2).

The lift force, normal to the flight path, is slightly less than in level flight because the component of weight W_2 opposite to lift is less than the actual weight.

For a powered aircraft in a descent, at constant airspeed and therefore constant drag, a small increase in thrust will enable a descend at a slightly reduced angle, reducing the magnitude of the weight component W_1 and reducing its propulsive effect. Drag is now balanced by the sum of the thrust and the small weight component. The aircraft is still in equilibrium at the same airspeed but with a reduced rate of descent. Further increases in thrust will enable smaller glide angles, eventually enabling level flight to be achieved with thrust equal to drag and lift equal to weight.

Lift, drag and the Resultant

In aerodynamics it is convenient to treat lift and drag separately. However, lift and drag are not really separate forces, they are components of a single force, the Resultant which is the sum of all of the separate aerodynamic forces acting on the aircraft. The forces in a glide can therefore also be represented simply by the weight acting vertically down and the resultant (the vector sum of lift and drag) acting vertically up. (Figure 3.3) This does not geometrically illustrate the angle of glide but the aircraft is still in equilibrium at constant speed and rate of descent. In this case we do not need to illustrate the drag component because it is included in the resultant.

Lift/drag ratio

3.4 Lift/drag Ratio

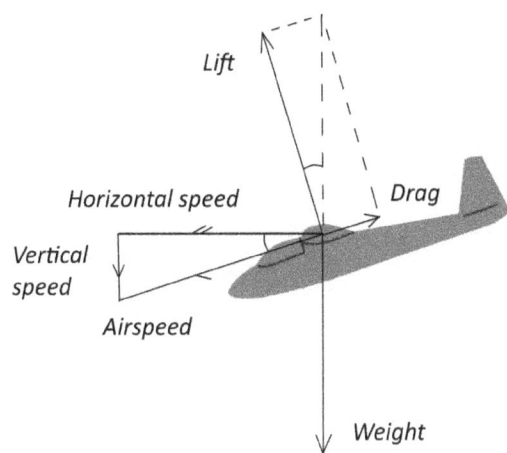

At any particular speed and angle of attack, there is a given relationship between lift and drag called the lift/drag ratio. Lift, drag and the resultant form a right-triangle. There is a similar triangle defining the glide-angle or glide-ratio of the aircraft. Therefore, the lift/drag ratio is the same as the glide ratio. (Figure 3.4).

In the case of the albatross, which has the best glide ratio of any bird with a lift/drag ratio of about 20, the glide angle is 3 degrees, which by coincidence is the same as the typical approach angle for aircraft landing at airports using the instrument landing system (ILS). However, although the angle of descent is the same, the airliners are not in a pure glide but rather they are making a powered descent at a given angle with the rate of descent controlled by the residual thrust.

3.3 Glider Equilibrium

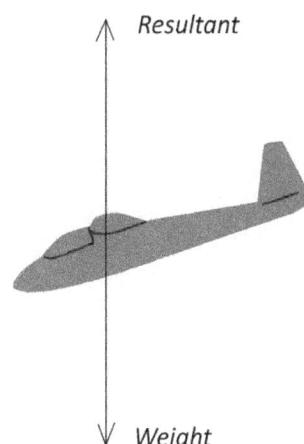

Lift/drag = glide ratio
= horizontal speed / vertical speed

Climbing

To gain height, thrust is increased to be greater than drag and the flight path is inclined upward. (Figure 3.5). Since lift is normal to the flight path but weight acts vertically downwards, we can again imagine the weight divided into two components; one component opposite to lift, and one component in-line with drag.

The increased thrust then equals the sum of the drag plus the component of weight W_1, at constant speed. The aircraft now has a constant rate of climb but note that the lift force is slightly less than in level flight because it is opposed by a component of the weight W_2 which is slightly less than the actual weight.

While excess lift will cause the aircraft to accelerate upward, when the aircraft starts to gain height, the direction of the relative airflow changes in a way that reduces the angle of attack and reduces lift. So, we do not increase lift in order to gain height. Only increased thrust will make the aircraft climb.

3.5 Forces in the climb

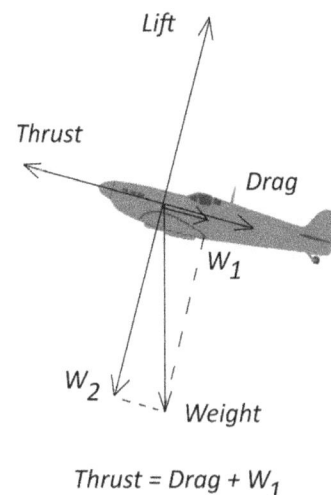

Thrust = Drag + W_1

Turning and load factor

Turning is achieved by banking the aircraft. The lift force tilts with the aircraft and there are then vertical and horizontal force components. The horizontal component acts as the centripetal force providing the centripetal acceleration which makes the aircraft turn. (Figure 3.6). Total lift is increased slightly, normally by increasing the angle of attack, so that the vertical component of lift equals the weight.

The increase of lift during turning flight compared with the lift in straight and level flight is called the load-factor. In straight flight, with the wings level, lift equals weight and the load factor is one (1G). In a level turn, the load-factor is equal to the reciprocal of the cosine of the angle of bank. At a 30 degree angle of bank, the load factor is 1.15G. At a 60 degree angle of bank, load-factor is 2G and so on. The need to increase lift during a turn results in an increase of drag, normally resulting in a loss of speed during turns. At small angles of bank up to about 30 degrees, a small loss of airspeed can be tolerated but at steeper angles of bank, over about 30 degrees, an increase of thrust is needed to prevent loss of airspeed. This is because the increased load-factor will cause an increase in stalling speed and a safety margin of speed needs to be maintained. A light training aircraft in a level 60 degree banked turn would normally need full-power to maintain speed and height compared to about 70% power in straight and level flight.

A glider in a turn needs to pitch-down a bit to prevent loss of airspeed with a consequent increase of its rate of descent. For a glider in a thermalling turn, a 45 degree angle of bank is usual, giving a reasonable compromise between a small turning radius and increased drag. A 60-degree steep turn would result in high drag and a high rate of descent.

3.6 Turning

Lift = weight x load factor

Vertical force = weight

Centripetal force

Weight

x = angle of bank

Load factor = 1/cos x

Diving

To recover from a dive, pitching-up causes an increased angle of attack, increased lift and increased load factor. The fact that the lift-force increases in a pitch-up or in a turn, means that the effective weight of the aircraft increases and therefore the stall speed increases. This means that an increased margin of airspeed is needed to avoid a stall at high G.

Stalling depends only on the angle of attack with the relative airflow and the aircraft can stall in any attitude relative to the ground. Although the stall always occurs at the same angle of attack, which corresponds to a certain aircraft pitch-attitude in level flight, this is less easy for the pilot to see when turning or pitching-up. To prevent stalling during aerobatics, in steep turns or while pulling out of a dive, the pilot has to be sensitive to pitch rate, rate of turn, control position and feel and airframe vibration known as buffet, caused by the impending stall.

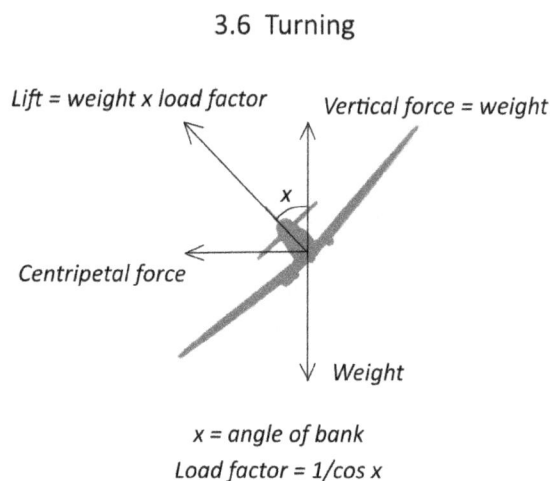

Aircraft axes

When an aircraft is manoeuvred, it rotates about its own axes, as in figure 3.7. For example, in straight and level flight, the aircraft rolls about its own longitudinal axis which is horizontal. However, if the aircraft is in a vertical dive, it still rolls around its longitudinal axis which is now vertical relative to the ground and its yaw and pitch axes are horizontal.

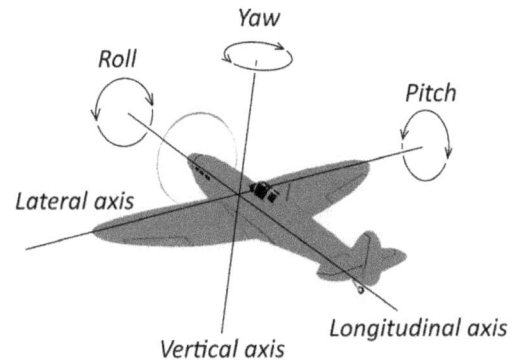

3.7 Aircraft axes

Aerodynamic equivalence

Aerodynamic forces are caused by the interaction of the airframe and the relative airflow. The same aerodynamic effects are produced whether the aircraft is moving and the air is stationary, as in free flight, or the aircraft is stationary and the air is moving, as in a wind tunnel. This is assuming that the aircraft is in equilibrium, that is, not under acceleration and that the relative air-velocity is uniform until deflected by the airframe.

Later we will look at dynamic soaring and see that the effects of unbalanced forces are to cause acceleration of the aircraft or bird. In this circumstance it *does* make a difference that it is the aircraft which has its own particular acceleration even though the wind is uniform. These effects are only seen when the aircraft or the bird is in free-flight and not held rigidly in a wind tunnel.

Angle of incidence

Angle of incidence, or rigger's angle of incidence, is the angle that the wing is attached to the fuselage. It is the angle between the wing chord line and the fuselage level line.

Wash-out

Stalling behaviour depends critically on whether the wing stalls first at the wing-root or the wing-tip. The former case is relatively benign and desirable. The latter is undesirable as typically one wing-tip stalls before the other and causes a wing-drop, which makes the recovery more difficult. This is dealt-with by building the wing with a slight twist which reduces the angle of incidence of the wing-tip compared with the wing-root. This is called wash-out. Alternatively, small stall-strips are added to the wing-root leading edge to trip the air-flow locally when close to the stall. This slightly increases the stall speed but is cheaper to manufacture compared to the complexity of a wing with wash-out.

Coanda effect

Air is a fluid and it is obvious that the solid airframe will deflect the air, with a degree of efficiency which will depend on the shape of the airframe. Less obvious is what happens when the air comes back together in the wake behind the aircraft. Up to a point, the fluid will follow the curved surface of the airframe, resulting in a deflection of the flow into the potential void downstream. This is known as the Coanda effect. Henri Coanda (1886-1972) was a Romanian engineer and inventor who worked throughout Europe and designed several aircraft for the Bristol aircraft company before WW1.

To demonstrate the Coanda effect, here is a neat experiment you can try: Lightly hold a spoon vertically between two fingers and run water over the convex surface of the bowl of the spoon. (Figure 3.8) You will see the water follow the curve of the surface and be deflected away from the vertical; and you will feel the spoon pulled towards the flow.

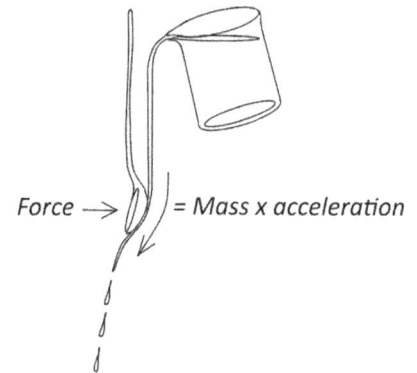

3.8 Coanda effect

Force → = Mass x acceleration

Initially, the fluid has vertical momentum due to gravity and has then been given horizontal momentum due to following the curved surface of the spoon. Momentum is mass times velocity. Rate of change of momentum is mass times acceleration. According to Newtons laws of motion, force equals mass times acceleration (F=m.a). In this case, the force acting on the spoon in one direction, is the equal and opposite reaction to the rate of horizontal momentum given to the fluid in the opposite direction. In the diagram the fluid is deflected to the left and the spoon is deflected to the right.

Aerodynamic force

The airflow around a wing responds to the Coanda effect and follows the curved upper and lower surfaces of the wing, producing a complex deflection of the airflow. The end result is a downwash in the wake of the wing, plus vortices shed by the wing tips. In other words, the relative motion of the wing and the air induces a rate of change of momentum of the air. Mass, in this case, is the air-mass being deflected by the wing at some rate according to the relative speed, angle of attack, air density and so on. The force acting on the wing is the equal and opposite effect of this rate of change of momentum in the air.

Newton

If we imagine an aircraft in straight and level flight, the momentum induced into the air by the wing, is partly downward and partly forward. The downward rate of change of momentum of the air results in the upward force known as Lift. The forward, horizontal momentum given to the air results in the rearward force known as Drag. This is the Newtonian explanation of lift in which force equals mass times acceleration. This explanation of lift suits our purpose when we later come to explain the transfer of momentum between the albatross and the wind during dynamic soaring. (Don't worry we will get to Bernoulli's theory shortly).

A wing, even a flat plate, with an angle of attack to the relative airflow, deflects the air and produces a downwash. The free-steam-velocity and the downwash velocity can be represented by two sides of a vector triangle. (Figure 3.9) The third side of the triangle represents the change of air-velocity caused by the wing; in other words, the velocity vector sum is the difference between the free-stream velocity and the downwash velocity. These vectors can represent magnitude and direction of velocity or of momentum (velocity times unit mass) or of rate of change of momentum (velocity times unit mass per unit time).

3.9 Velocity, momentum and force vectors

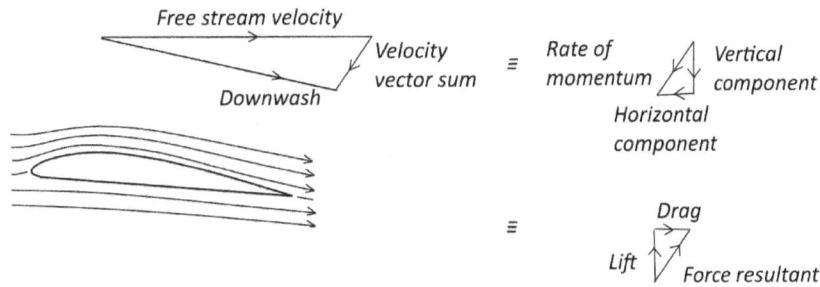

The velocity vector sum is equivalent to the rate of change of momentum and can be divided into two components of vertical momentum and horizontal momentum, orthogonal to the free stream velocity. Each of these components, corresponds to a vertical lift force and a horizontal drag force. Lift and drag forces can then be represented by the single Resultant force. Thus, the velocity vector sum is what causes the Resultant force, which comprises Lift and Drag force components.

Ground effect

This is a good moment at which to mention ground effect, which is a characteristic of albatross dynamic soaring. Compare figures 3.9 and 3.10. When the wing is flying within about half a wing-span of the surface, it can produce the same lift at slightly reduced angle of attack. Thus, with a slightly reduced angle of attack, the down-wash angle is reduced and the vector-sum is modified. The vertical component of momentum is unchanged and the lift force remains equal to the weight, while the forward component of momentum given to the air is reduced and therefore the drag force is reduced.

3.10 Ground effect

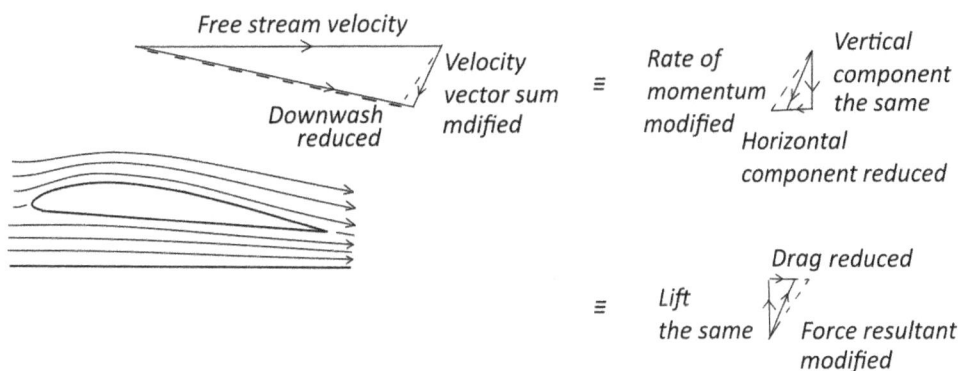

Bernoulli

Bernoulli's principle is equally valid and explains the pressure distribution around the wing. Bernoulli's principle says that the sum of static and dynamic pressure in a moving fluid is constant.

Static pressure is ambient or atmospheric pressure. Dynamic pressure is the pressure caused when fluid is accelerated. So for example when you hold your hand out of a moving vehicle, the air impacting your hand is slowed down and you feel a force acting on your hand.

At sub-sonic speeds air behaves as if it is incompressible. In a pipe containing moving fluid, if the fluid is forced to change speed due to a constriction in the pipe, the change in the dynamic pressure is balanced by an opposite change in static pressure. As the fluid speeds up in the constriction, the dynamic pressure increases and the static pressure reduces. (Figure 3.11). Downstream the static pressure returns to normal.

3.11 Pressure in a venturi

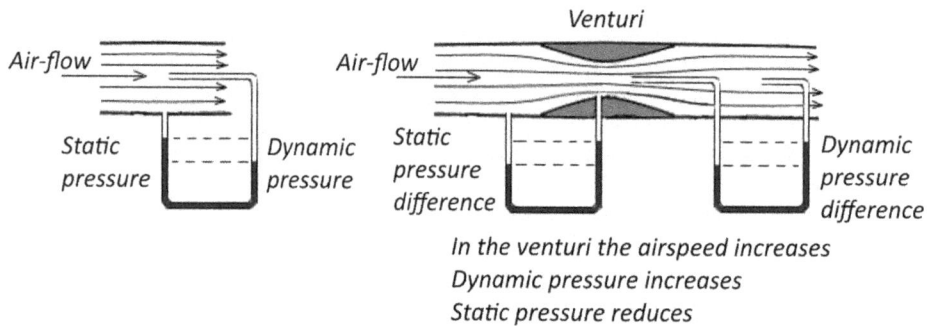

In the venturi the airspeed increases
Dynamic pressure increases
Static pressure reduces

In the case of an aircraft wing, if the wing is cambered or there is an angle of attack, the interaction of the wing and the relative airflow causes the air to travel faster over the top of the wing compared with under the wing. This results in the air pressure being less above the wing compared with under the wing and this accounts for the lift force. It is the same effect as previously explained in Newtonian terms. This is illustrated in figure 3.12, with low pressure above the wing and high pressure below the wing, along with the high pressure at the leading edge which is the stagnation point.

3.12 Pressure distribution

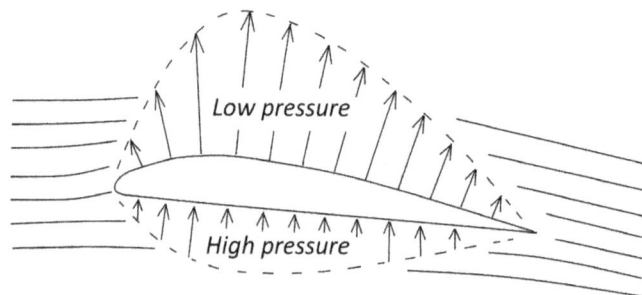

How does Bernoulli's theorem account for drag? If you imagine the air pressure acting normal to each small increment of the wing surface, you can see that most of the upper and lower surfaces will tilt the action of the pressure force slightly in the downstream direction. Thus the pressure distribution causes both lift and drag.

Stalling

The lift force increases as the angle of attack increases up to about 15 degrees, the wing working at its greatest lift coefficient but with high drag.

At the trailing edge, the difference between the low pressure above the wing and the high pressure below the wing, makes the air try to curl around the trailing edge. This is resisted by the momentum of the airflow up to about 15 degrees angle of attack, at which point the pressure gradient wins out and the airflow breaks away from the wing into a detached turbulent flow. The lift reduces, the drag increases and this is the stall. (Figure 3.13).

In flight training, stalling is typically demonstrated in straight and level flight. The approaching stall is seen as a high nose-attitude, reducing airspeed and sloppy controls and is felt as airframe buffeting. The actual stall is seen as a nose or wing drop; the stalling airspeed can be seen on the airspeed indicator. Recovery from the stall is achieved by pitching down and reducing the angle of attack; at which point the airflow re-attaches and becomes smooth once again and the airspeed increases to normal in a dive or under power.

3.13 Boundary layers and stalling

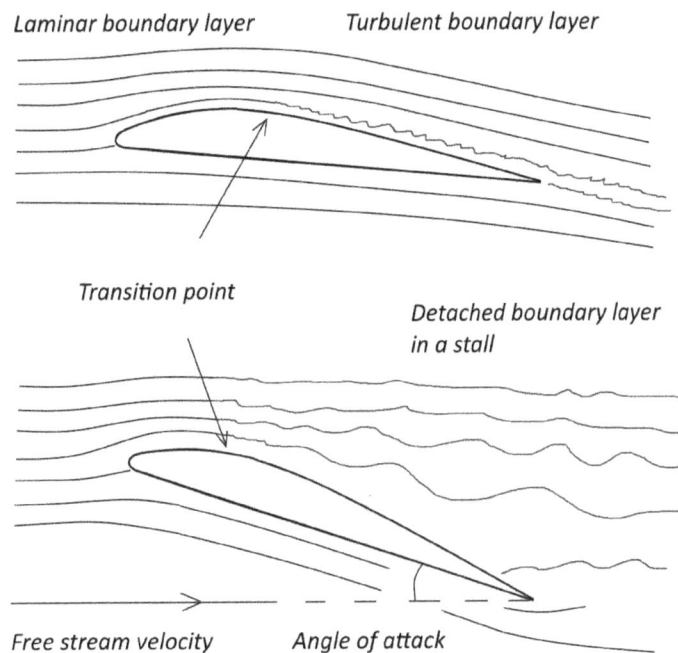

Laminar boundary layer   Turbulent boundary layer

Transition point

Detached boundary layer
in a stall

Free stream velocity   Angle of attack

Stalling speed

Stalling depends on angle of attack and not on airspeed; the wing can stall in any aircraft attitude, in a climb, a dive or a turn and at any speed. However, since lift depends on angle of attack *and* airspeed, large angles of attack are associated with slow airspeeds and stalling is referenced to stalling speed in un-accelerated (1G) flight. Stalling speed is the airspeed at which the wing reaches the stalling angle of attack at any particular load factor.

Stalling speeds increase with increased weight but can be reduced by using devices such as trailing edge flaps and leading edge slots, at the cost of increased drag. That cost is not a problem if the devices are only used at slow speeds and are retractable for flight at normal speeds.

Boundary layer

Air is a fluid and, despite its low density, it behaves similarly to other fluids like water. Specifically, the airflow forms a thin boundary layer around the airframe. The boundary layer can be thought of as a number of very thin layers of fluid; the layer actually touching the airframe skin is stationary relative to the surface and each successive layer further from the surface moves a little faster, until the outer layer of the boundary layer is moving at the full fluid velocity. It is difficult to define the thickness of the boundary layer but one way is to say that the outer-most layer is where the fluid velocity has reached 99% of the free stream velocity. Within the boundary layer, viscosity causes each infinitesimally thin layer of fluid to drag on the next, causing the effect of skin friction.

Laminar and turbulent flow

The attached boundary layer comes in two flavours: laminar and turbulent. Either sort can then become detached from the surface. The laminar boundary layer is thin and low-drag and is typically found at the leading-edge of the wing and the front of the fuselage. At a certain point downstream, the transition point, the laminar boundary layer structure breaks-down into a turbulent flow and the turbulent boundary layer becomes thicker but remains attached to the skin. Compared to laminar flow, in the turbulent boundary-layer the skin-friction drag is greater due to the greater micro-vorticity induced in the air but the boundary layer still follows the wing profile until close to the trailing-edge.

As the angle of attack increases, a laminar boundary layer is more likely to break-down into a detached flow, compared with a turbulent boundary layer which remains attached to the surface to a greater angle of attack. A laminar flow wing section is also more sensitive to contamination. For example, dead bugs or ice on the leading edge will also cause the laminar boundary layer to trip into turbulent flow.

When designing wing-sections for aircraft wings, an extensive laminar boundary layer is desirable to reduce drag at small angles of attack, that is to say in cruising flight. However, this can lead to a sudden break-down of the boundary layer at the stall and cause rather unforgiving stalling characteristics; as compared to the more progressive break-down of a turbulent boundary layer. So, a training aircraft would be designed with a relatively blunt air-foil, giving a turbulent boundary layer with minimal laminar flow; to give it good handling characteristics at the stall, at the cost of greater skin-friction drag. Whereas, a high-performance aircraft would be designed with a thinner wing, a sharper leading-edge and more extensive laminar flow for least drag at cruising speeds; which may have less-benign stalling characteristics but will be fitted with aids for the pilot to help prevent stalling. Modern sailplanes are built with laminar flow wing sections with relatively sharp leading edges to maximise their lift/drag ratio. However, they are notorious for losing their laminar flow due to water droplets or dead bugs on the leading edges and jumping suddenly into a high-drag semi-stalled condition even at low angles of attack!

Aeromodellers have found that model aircraft, which are more like the size of birds than the size of manned aircraft, work best with a turbulent boundary layer. It has been found worthwhile to add turbulators to the leading edges of model gliders to trip the laminar plus semi-detached boundary layer into a turbulent but fully-attached boundary layer and so reduce the overall drag.

It seems likely that birds have turbulent boundary layers to reduce drag. Feathered surfaces are smoothly contoured but relatively rough compared with smooth polished artificial surfaces. Other flying animals like bats and insects also have quite rough surfaces.

Drag

Drag is the force component opposite to the motion of the aircraft. The drag force is the equal and opposite effect to the rate of change of momentum induced in the air. Drag forces are of four types:

1. Form drag is caused by the displacement of the air by the passing airframe, in other words the air is simply pushed aside. Drag is the force felt by the airframe as the reaction to the rate of momentum given to the air as it is displaced.

2. Interference drag is caused by the various deflected air-streams conflicting with each other and causing vorticity in the air, for example where the wing joins the fuselage. Vorticity is rotational motion and is another form of momentum in the air.

3. Skin friction is caused by the viscous shear forces within the boundary layer adjacent to the skin of the airframe.

4. Induced drag or, more properly lift-induced drag, is caused by the pressure difference between the top and bottom of the wing. This causes the air to spill around the wing-tip from high pressure below the wing to low pressure above the wing, creating a spinning vortex behind each wing-tip. These vortices comprise rotational momentum which is left behind and dissipated in the atmosphere. (Figure 3.14). The deflected air at the wing-tips generates span-wise components in the airflows above and below the wing, deflecting the airstream inward above the wing and outward below the wing. These two air-flows meet at the trailing edge of the wing, moving in slightly different

3.14 Wing-tip vortices

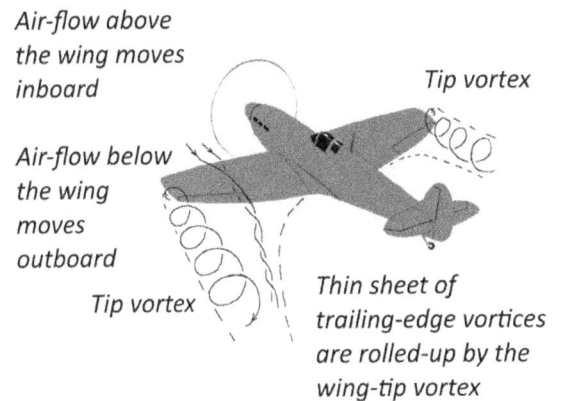

Air-flow above the wing moves inboard

Tip vortex

Air-flow below the wing moves outboard

Tip vortex

Thin sheet of trailing-edge vortices are rolled-up by the wing-tip vortex

directions and cause a thin sheet of small vortices behind the wing which is then gathered up by the wing-tip vortices to create two large parallel vortex tubes trailing behind the wing-tips.

Parasite drag is caused by the various excrescences attached to the airframe like aerials and undercarriage legs. It is caused by combinations of the above four types of drag.

Drag caused by supersonic shock waves is another issue but beyond the scope of this book.

Minimising drag is a preoccupation of aircraft designers, although you might not think so from looking at some of the products of the industry, particularly in the biplane era. In practice there is a trade-off between, on the one hand, minimising drag and running costs and, on the other hand, reducing weight and production costs.

Starting vortex

When the aircraft begins to move at the beginning of the take-off, a starting-vortex is formed along the whole trailing-edge due the air under the wing curling around the trailing-edge. As the momentum of the flow over the upper surface builds up, the trailing-edge vortex is shed from the wing and left behind on the runway but is joined to the wing-tip vortices which continue in flight. (Figure 3.15).

3.15 Starting vortex

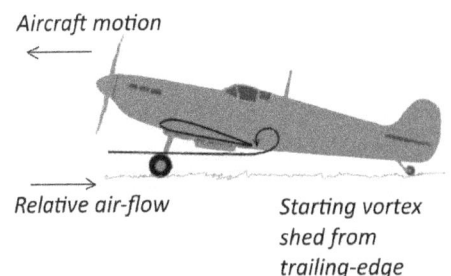

Aircraft motion

Relative air-flow

Starting vortex shed from trailing-edge

Stopping vortex

The formation of the wing-tip vortices continues from take-off to landing. After landing, when the wing stops producing lift, the wing-tip vortices cease and a vortex is shed from the leading-edge. That is to say that the wing-tip vortices detach from the wing, along with the leading edge vortex and drift away, completing the whole vortex ring, along with the starting vortex which was left on the departure runway. (Figure 3.16).

3.16 Stopping vortex

As the aircraft slows after landing a leading-edge vortex, which is connected to the wing-tip vortices, is formed and drifts away

The drag curve

Drag comprises principally form drag and induced drag. Form drag increases with the square of the airspeed and is greatest at high speed, whereas induced drag is inversely proportional to the airspeed, being greatest at low speeds and high angles of attack.

When the effects of form drag and induced drag are plotted on a graph of drag versus airspeed, the sum of the two drag curves gives a U-shaped total drag curve. (Figure 3.17).

3.17 The drag curve

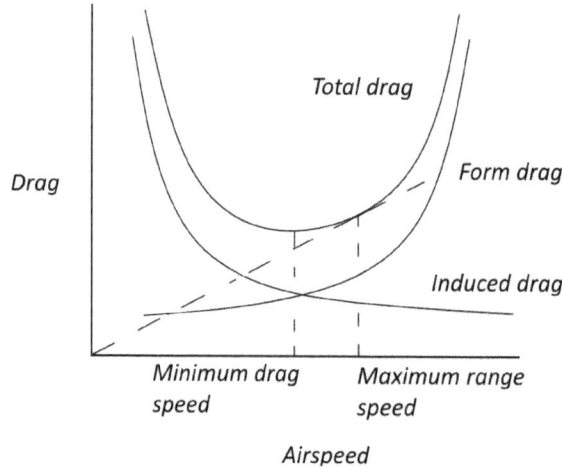

Minimum drag speed

The minimum drag is seen at the lowest point of the U and gives the airspeed to fly for maximum duration. For a powered aircraft, the least drag will mean the least fuel consumption and the maximum endurance. For a glider, the minimum drag will give the airspeed for minimum rate of descent (minimum sink speed) and therefore the maximum flight time for any given height loss.

Maximum range speed

A line drawn from the origin of the graph touching the total-drag curve at a tangent gives the airspeed for maximum range, an airspeed slightly greater than the minimum drag speed. In other words, it gives the maximum ratio of airspeed to drag. For a powered aircraft, the maximum range speed gives the maximum range for a given amount of fuel consumption in still air. For a glider it gives the speed for best glide angle, which gives the maximum distance flown for a given height loss, again in still air.

Effect of the wind

For the maximum range speed, a headwind or tailwind gives a different result. Clearly, a headwind equal to the airspeed would give zero groundspeed and result in complete exhaustion of fuel or height without any distance travelled at all. To get the best range speed in a wind, the origin of the tangent line is drawn from the headwind or tailwind position. For a headwind, drawing the tangent from the headwind gives a steeper line and a greater airspeed for maximum range. The maximum range in a headwind is, of course, less than the maximum range in still air. Doing the same thing for a tailwind yields a lesser airspeed for maximum range but in practice there is little benefit. For gliders the same principle applies and the best glide speed is greater in a headwind but the maximum distance achieved from a given height is less than in still air.

The lift curve

Aircraft normally fly straight and level in 1G flight with lift equal to weight; they fly at different angles of attack at different airspeeds to maintain constant lift. Therefore, rather than plotting lift against speed, it makes sense to plot lift coefficient against angle of attack. (Figure 3.18).

3.18 The Lift curve

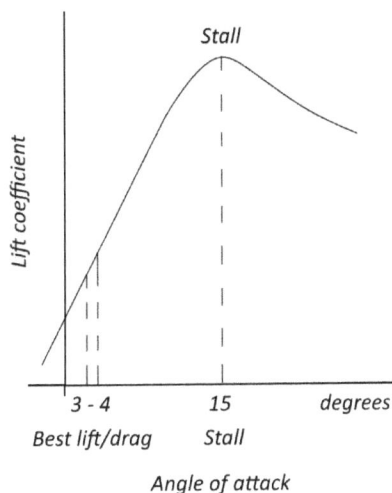

The lift curve gives a fairly straight line up to about a 15 degree angle of attack and then a reduction of lift indicating the stall. Although there is still a lot of lift above the stall, this corresponds to the very high drag condition at low airspeed, as seen on the drag curve, and is therefore very inefficient.

In normal flight, stalling is avoided and the wing is made to work at small angles of attack, around 3 to 4 degrees for the best ratio of lift to drag. Therefore, the wing is attached to the fuselage at this angle of incidence, known as the riggers angle of incidence, so that the fuselage is effectively at zero angle of attack for minimum drag at cruising speed.

Polar curves

For glider pilots, the aircraft performance is presented on a polar curve which is a plot of sink rate (negative rate of climb) against airspeed. (Figure 3.19). Here, the minimum sink speed is the airspeed at the apex of the curve while the best glide speed is the airspeed indicated by the tangent of the line from the origin. Once again, in a headwind the speed to fly for maximum distance over the ground is greater and is given by the tangent of a line drawn from the wind-speed.

3.19 The glide polar curve

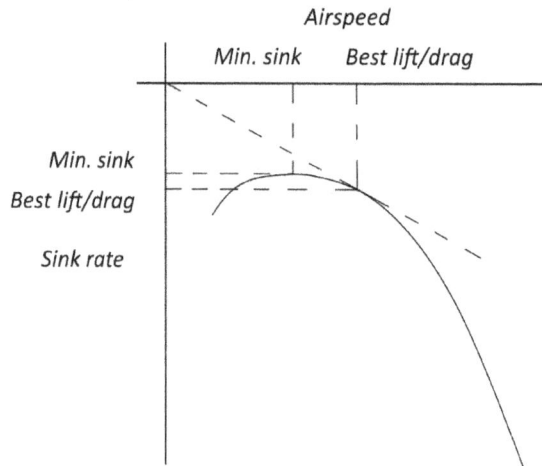

Control and Stability

The conflicting requirements of control and stability are fundamental to aircraft design. An airliner needs a high degree of stability and relatively modest control authority to give the passengers the smoothest ride. A sport aircraft needs much less stability but greater manoeuvrability with light yet powerful controls.

Directional stability

In aircraft, lift and drag act at the same point known as the centre of pressure (CoP). Stability is achieved by the relative disposition of the CoP and the centre of mass. Dart-like directional stability is achieved by having the centre of mass ahead of the CoP. In the case of an arrow, with a heavy tip, the CoP, dominated by the vanes at the back, is a long way behind the centre of mass. The inertia of the dart or arrow takes it on a ballistic trajectory with the stabilising fins following in-line. Any deflection of the arrow away from its line of flight results in an angle of attack on the tail fins and a sideways force to push the tail back in-line.

3.20 Cayley glider 1904

The very first aircraft built by George Cayley in 1804 was a glider which combined a dart and a kite. The kite, set at a small angle of incidence provided the lift and the dart fins at the back, combined with a weight at the front, gave the aircraft stability. (Figure 3.20).

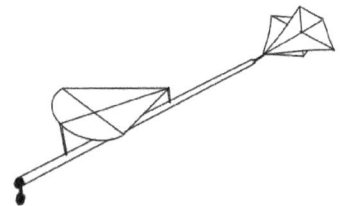

Pitch control

Pitch is rotation about the aircraft lateral axis, the wing-tip to wing-tip axis, meaning that the aircraft nose moves up and down. Control of the aircraft in pitch is achieved by moving the control stick forward or back, deflecting the elevator down or up. Some aircraft have an all-moving tailplane combining the function of tailplane and elevator. When flying straight and level at different speeds, different angles of attack are required and thus different pitch attitudes. At low speeds, a large angle of attack and a high nose attitude is needed.

Pitch Stability

The best arrangement for the pitch stability of an aircraft is to have the centre of mass ahead of the centre of drag, as it is with a dart. Meanwhile, the lift-force must act close to the centre of mass so that the lift force is acting opposite to the weight. The centre of lift acts at the same place as the centre of drag, known as the centre of pressure, which means that the lift force is also slightly behind the centre of mass and causes a nose-down moment. To balance this, the tailplane provides a down-force and a nose-up moment. Therefore, the centre of mass is made to be close to, but slightly ahead of, the centre of pressure, which means that the balance of the aircraft is a critical matter to maintain longitudinal stability. (Figure 3.21).

3.21 Pitch stability

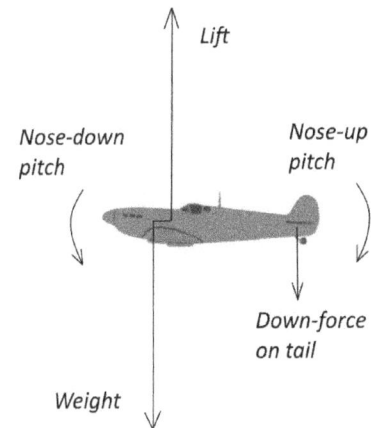

This arrangement of the tailplane at a small negative angle of attack and the wing at a small positive angle of attack is called longitudinal dihedral. The aerodynamic load produced by the tailplane working at a slightly negative angle of attack means that the price of stability is a small drag penalty, both because of the need for a surface at the tail but also because the wing must produce more lift to support the additional download on the tail. The drag penalty is minimised by the fact that both surfaces are working close to their best lift/drag ratio angle of attack, about 2 to 4 degrees. The tailplane also produces dart-like stability.

Stability can also be achieved in canard-type tail-first aircraft by having the stabiliser at the front with a positive angle of attack greater than the wing. However, this arrangement means that neither surface is working at its best lift/drag ratio angle of attack and therefore this is not as efficient as having the stabiliser at the rear.

Yaw control

Yaw is rotation of the aircraft about its own vertical axis. An aircraft normally has a vertical fin for yaw (directional) stability and a rudder for yaw control. The rudder is controlled with the pilot's feet. Pushing the left pedal gives left rudder and vice versa. Deflecting the rudder will cause the aircraft to yaw in that direction, which is the primary effect. (Figure 3.22).

3.22 Primary and secondary effects of yaw

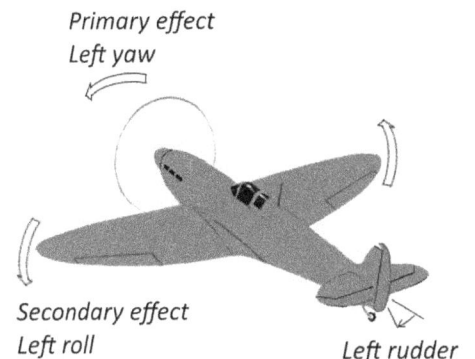

The secondary effect of rudder is that the aircraft will roll in the same direction as the yaw. This is because the yaw motion causes the outer wing to move faster and, if the aircraft has dihedral, the outer wing will gain a greater angle of attack, causing that wing to produce more lift. This rudder-only form of control is not normally used for turning manned aircraft, although simple radio control aircraft, with no aileron control but with a large dihedral angle on the wings, can be controlled in this way.

Yaw stability

The fin will keep the aircraft pointing into the relative airflow if the aircraft is disturbed about its yaw axis. However, it will tend to destabilise the aircraft if the aircraft is disturbed into a banked attitude and starts to side-slip towards the lower wing. Then, the fin will yaw the aircraft towards the lower wing which leads to more roll and then more yaw and the aircraft drops into a spiral dive. Thus, strong yaw stability can lead to reduced roll stability. More on this later.

Yaw stability can also be achieved by using swept back wings or upturned wing-tips and this is done with some tail-less aircraft. When a swept-wing aircraft yaws, the wing moving forward effectively has less sweep and more span while the wing moving back has more sweep and less span. Thus the drag forces become asymmetric and restore the aircraft back in line with the relative airflow. Up-turned wing tips simply act like short-coupled fins, being only a short distance behind the centre of mass despite being out on the wing-tips. So-called flying-wing aircraft are notoriously weak in directional stability and often require fins for yaw-damping and drag surfaces at the wing tips for yaw control.

Roll control

Roll is motion about the aircraft fore and aft or longitudinal axis. Roll control is typically achieved with ailerons as follows: To roll right, the control stick or wheel is moved right, the right aileron goes up and the left aileron goes down. This reduces the lift on the right wing and increases the lift on the left wing and the aircraft rolls right. That is the primary effect. (Figure 3.23).

At the same time, the secondary effect is that the drag on the right (down-going) wing reduces and the drag on the left (up-going) wing increases, causing the aircraft to yaw to the left, opposite to the roll. This is known as adverse yaw or

3.23 Primary and secondary effects of roll

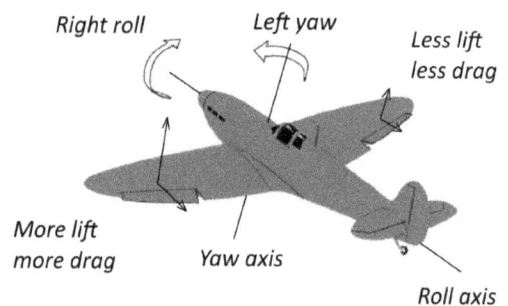

Right roll   *Left yaw*
Less lift less drag

More lift more drag   *Yaw axis*

Roll axis

aileron drag. This is a particular problem for sailplanes with long, narrow high aspect-ratio wings and for regular aircraft at low speeds and high angles of attack. Less so for aircraft with small wings at higher speeds. To cancel the adverse yaw, the pilot must apply a small amount of rudder in the same direction as the turn at the same time as rolling into the turn.

Before the invention of ailerons, the Wrights and other early designers used wing-warping for roll control in which the whole wing was twisted to achieve the change of incidence. Some pioneers thought that the rudder should be the main turning control as with a boat and that the warping control should be used to prevent roll and keep the wings level. Wing warping was not very effective as it was often found that the secondary yaw was greater than the primary roll effect leading to control reversal. Also, it led to very flexible airframes and poor control response.

Wing-warping was superseded by ailerons which were much more effective but still caused adverse yaw until innovations like differential ailerons and Frise ailerons were introduced. With differential ailerons the up-going aileron deflects more than the down-going one to reduce the adverse yaw but this also reduces the effectiveness of the ailerons. Frise ailerons work by having inset hinges so that the leading edge of the up-going aileron protrudes below the wing to increase the drag on the down-going wing-tip.

When the aircraft starts to roll, the motion of the down-going wing gives it a higher angle of attack and greater lift which resists the roll motion. This is known as roll damping.

Roll stability

Roll stability is achieved in several ways; by having high-mounted or shoulder-mounted wings to give pendulum-like stability or the wings can be set at a dihedral angle or with sweep back. Dihedral is intended to correct for a wing-down condition with or without side-slip. The lift force on each wing is normal to the wing surface so that, as soon as the aircraft with dihedral is disturbed in roll, the lower wing's lift vector becomes more vertical and restores the aircraft to the wings-level condition. Additionally, as the aircraft rolls, it starts to descend and side-slips towards the lower wing. The increased angle of attack on that side increases the lift and raises the wing.

3.24 Dihedral and side-slip

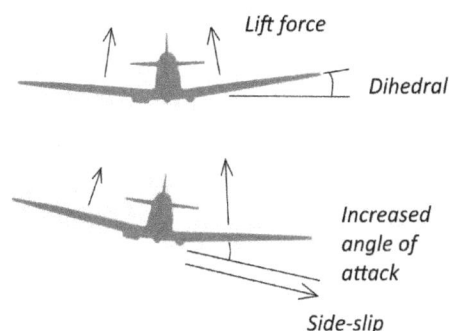

Lift force

Dihedral

Increased angle of attack

Side-slip

Spiral instability

The side-effect of the banked attitude in a side-slip, is that the increased lift on the lower wing means that there is increased drag which tends to pull the wing back and yaw the aircraft towards the lower wing, opposite to the recovery roll. In other words, dihedral or sweep will pick-up the wing in a side-slip but will cause yaw in the direction of the lower wing, leading to a spiral dive again. In a banked side-slip the fin will try to yaw the aircraft towards the lower wing causing more roll in that direction and leading once more to a spiral dive. Thus you can see that the requirements for roll and yaw stability are at odds with each other, strong roll stability leads to poor yaw stability while strong yaw stability causes poor roll stability.

For small aircraft with manual controls, the design preference is towards strong directional (yaw) stability with less lateral (roll) stability, but giving lighter aileron controls, making it relatively easy to keep the wings level.

Dutch Roll

Big, fast aircraft utilise powered controls so that control forces are not a problem for the pilot. The aircraft can be designed with dihedral for strong lateral stability and sweep-back for both stability and higher critical Mach numbers, enabling jet aircraft to fly closer to the speed of sound. For large aircraft, designers prefer to reduce fin and rudder areas to reduce drag, however, this can cause a problem called dutch-roll (so-called after the swaying gait of ice-skaters).

Dutch-roll is an out-of-phase roll-yaw couple; in other words, the aircraft alternately rolls one way while yawing the other way. This is due to a combination of strong roll stability, caused by dihedral and swept-back wings, combined with poor directional stability due to a small fin and rudder. It is exacerbated by a large moment of inertia due to masses at a distance from the centre of mass, for example engines mounted at the tail or outboard on the wings.

The cure for Dutch-roll is normally achieved with a yaw-damper in the rudder-control circuit. This prevents the rudder from deflecting due to the side-slip angle, effectively increasing the fin area and reinforcing yaw stability, while still allowing the rudder to act as a yaw control.

Hang Gliders

You may well ask why hang-gliders don't need fins and rudders? The pilot of a hang-glider is suspended in a harness below the glider's centre of lift and steers by bracing himself against an A-frame; shifting his weight left or right for steering and forwards or backwards for pitch control, to go faster and slower.

What happens when the pilot wants to turn? He shifts his weight to the side and the aircraft rolls in that direction. The down-going motion of the wing gives the wing a greater angle of attack on that side and paradoxically more lift and drag than the up-going wing. This lift differential acts as a roll-damping force-couple and the drag differential gives a yaw in the same direction as the roll.

Circulation

There is another way of describing how lift and drag are caused. Aerodynamic effects can be described as a combination of the free airstream velocity and an assumed circulation flow around the wing. This is the system used in computational fluid dynamics as it is relatively easier to translate into computer code. The air does not actually circulate around the wing; it definitely flows from front to back. However, within the overall airflow there are various degrees of rotation. (Figure 3.25).

The airflow around the wing has a vertically upward component *a* ahead of the wing. Then, above the wing there is a horizontal rearward acceleration component (speeding-up) *b* and then a forward acceleration component (slowing-down) *c*. These two components sum up to give a small speed-up *f* above the wing. Next, there is a downward component *d* behind the wing. Meanwhile under the wing there is a horizontal forward component of acceleration *e* as the air-flow here slows-down. These five components, when added up, form a notional circulation around the wing called a bound vortex which accounts for the downward and forward motion of the airflow, vector sum *v*, which we saw earlier in figure 3.9. Combined with the free airstream velocity, this results in the overall downward

3.25 Circulation

deflected and slowed airstream behind the wing. This change of velocity over time is equivalent to a rate of change of momentum in the air, the equal and opposite effect of which is the aerodynamic resultant force acting on the wing. The resultant then gives the lift and drag force components.

Vortex theory

According to vortex theory, a vortex cannot end within a fluid but only at the boundary of the fluid; which would be a solid surface or a boundary caused by a marked density change like the surface of the sea.

Hence, the circulation around the wing forms a bound vortex which is part of a larger joined-up vortex ring comprising, the wing-tip trailing vortices, the starting vortex and the stopping vortex. Once detached from the wing this vorticity will dissipate over time but can still affect an aircraft following several miles behind.

When the wind blows

For an aircraft in equilibrium, airspeed is maintained by a balance of thrust and drag. The effect of the wind is to generate an angle of drift and a ground-velocity which is the vector sum of air-velocity and wind-velocity. (see chapter 7 Navigation). When the wind is changing from one state of equilibrium to another, the air is accelerated and the aircraft air-velocity will change by a small amount (as in turbulence) causing changes to the aerodynamic forces, specifically drag. The changing drag-force then causes the ground-velocity to change, until a new state of equilibrium is achieved and the aircraft has returned to its original airspeed with the drag balanced by thrust. For example, if the headwind increases, airspeed increases slightly while ground-speed is initially unchanged due to inertia resisting any change of ground-velocity. As drag increases, ground-speed and airspeed reduce together and a new state of equilibrium is reached with the original airspeed but with reduced ground-speed corresponding to the new headwind component.

Turbulence

In turbulence, sudden, brief accelerations of the air result in corresponding changes to air-velocity, causing fluctuations of airspeed and aerodynamic forces, while changes of ground-velocity (acceleration of the aircraft) are resisted by inertia. Thus the turbulence is felt as bumps or air-pockets, as some people like to call them.

Acceleration

Airspeed is relative speed and is affected both by acceleration of the wind and by acceleration of the aircraft. Either acceleration can be either centripetal or tangential, so that changing headwind-components can be due to changing wind-velocity or due to changing aircraft-velocity relative to the wind, which could be due to changing wind-angle or changing heading.

Consider, for example a winch-launched glider. During launch, the machine is lined-up facing the wind and accelerated by the winch-cable causing its groundspeed to increase. At the same time as the glider is accelerating, imagine that the wind itself is increasing, causing an increasing headwind. The airspeed of the glider will be the sum of the ground-speed and the headwind component. The acceleration of airspeed will be the sum of the acceleration of the ground-speed and the acceleration of the headwind component. Hold that idea, we will come back to it later.

Chapter 4

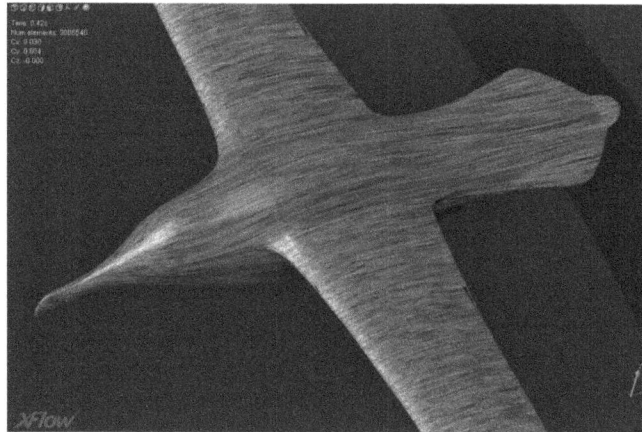

Theory of flight in birds

Anatomy

Among the 10,000 or so species of birds the majority are flying animals and while they exhibit a wide range of sizes and plumage from wrens to vultures and from sparrows to pheasants, nevertheless there is little variation in their anatomy. Even including the non-flying birds, the variation is not great; the penguin and the ostrich being the most extreme. In comparison, the variation of form and size among the mammals is considerable; with the bats, the whales, the shrews and the elephants being at the various corners of the evolutionary envelope.

The driving force controlling the form of birds is, of course, flight. All birds are evolved from a relatively small number of flying birds that survived the mass extinction which killed off the dinosaurs at the end of the Cretaceous era 65 million years ago. Every aspect of bird anatomy is modified by the needs of flight to achieve lightness combined with strength. Most notably their bones are substantially hollow and contain a web of fine supporting struts together with air-sacs which form part of their respiratory and cooling system. Compared with the dinosaurs from which they are most likely evolved, birds have lost the long bony tail, the toothed jaw and the clawed hand and they have acquired a lighter beak made of keratin, a fused hand structure and of course, flight feathers. The thoracic and pelvic structures have become a relatively rigid box of bony struts with a well-developed sternum or keel to support the flight loads and associated muscular stresses. The respiratory system relies on a system of air sacs manipulated by an expandable rib cage rather than a diaphragm, to promote airflow through the lungs.

Feathers

Birds differ from aircraft in that they have soft bodies covered by feathers rather than a hard outer skin. It has been established by experimentation that live birds have less drag than model birds of the same size and shape. Newly hatched chicks, having lost their long bony dinosaur tails to evolution, have conspicuously blunt rear ends but, once they have grown their flight plumage, they are covered by contour feathers forming a streamlined body, with the flat tail feathers available for control purposes.

The feathers enable the body to reduce drag by conforming to the airflow, at different speeds and angles-of-attack, rather than having a rigid body trying to force the airflow to conform to its contours, which could only work efficiently at one speed and angle-of-attack. Watching bird feathers in flight, the airflow can be visualised to some extent because the feathers lie flat where the airflow is smooth, while they lift and flutter where the airflow is turbulent or detached. Every feather has both muscle fibres and nerve endings at its insertion point so the bird is undoubtedly aware of any turbulence around its body as it affects each individual feather. It is, to some extent, able to control each feather, particularly by applying torque loads on the feather's long axis, to minimise air leakage between feathers and by aligning it with the airflow. This creates an active feed-back mechanism to reduce drag.

The feather tips at the trailing edge of the wing are very thin and flexible and are thus able to conform to the airflow and reduce turbulence and drag. The pointed feather tips form a serrated wing trailing-edge which will allow smoother mixing of the airflows from the upper and lower surfaces. This has been adopted in full-size aircraft design and is seen at the rear edges of the cowlings of big-fan turbo-jet engines where the high speed air from inside the cowling mixes with the slower speed air outside the cowling. This improves efficiency and reduces noise, which, of course, owls have been doing for millennia.

Some aircraft are fitted with fixed or moveable, slotted wing leading-edges to energise the airflow over the top surface and delay the stall, in order to improve handling at low speeds. Birds have a similar device in the form of a feather called the alula attached to the vestigial thumb part of the hand-wing. This will generate a vortex and re-energise the airflow over the top of the wing at high angles of attack.

In terms of their size, birds are more like model aircraft than full-size aircraft and therefore operate at similar Reynolds numbers. The Reynolds number is a non-dimensional number indicating the relative significance of viscous and inertial fluid effects. Viscous effects are more significant at small sizes and inertial or rotational effects at larger sizes. Model aircraft designers have found that small aircraft work best with a turbulent boundary layer to minimise overall drag. It is found that sharp leading edges and rough surfaces maintain attached, turbulent boundary layers and minimise detached flows at the cost of slightly increased skin friction. Shiny surfaces on model aircraft may look good but do not work so well aerodynamically. Therefore, it seems unlikely that bird wings have much laminar flow; although they do seem to have different degrees of roughness in their plumage, for example above and below the wings.

Control and stability

When pilots are taught to fly, they are given particular rules to follow in order to sort out their control inputs. For example, on the approach, airspeed is controlled by pitch attitude and elevator control, whilst rate of descent is controlled by engine rpm or by air-brake. However, when watching an experienced pilot making an approach, all of the control inputs are smoothly blended and coordinated so that it is difficult to see which control is achieving which action. So it is with birds, without knowing the bird's intention, all of the wing and body movements are smoothly blended to achieve a particular outcome, which is not known by the observer until after it is completed!

Birds are of course, powered aircraft when they are flapping and are gliders when they are not flapping. Unlike manned aircraft, which have separate surfaces for thrust, lift and control, birds combine these three functions into a single aerodynamic system, more like a helicopter. So, even when a bird is gliding, very small movements of their wings could contribute to drag reduction, thrust and control. Like helicopters, birds are also likely to be marginally stable and their control deflections are going to be so small as to be almost imperceptible.

Birds can get away with more or less neutral stability because their senses, instincts and reflexes are so finely tuned by evolution and their whole body is their control system. Yet this is not so different from the experience of the average human being who balances and stands on two feet and runs and jumps more or less automatically. When a person wants to stand still with minimum effort, they will lock-out their knee joints and arrange their posture so that their centre of mass is over the contact patch that their feet make on the ground. When they want to walk or run they lean forward so that their centre of mass is outside of the contact patch and they become unstable. So it must be with birds. Their aerodynamic centre of pressure must coincide with their centre of mass but will move around slightly to achieve an appropriate balance of control and stability according to the circumstances. Slightly more stability will reduce the effort to maintain equilibrium. Slightly less stability will reduce the effort for manoeuvring. Locking elbow joints will reduce the effort to keep the wings extended.

Pitch control

How do birds deal with pitch control and stability? They can do this in two ways. Firstly, birds can swing their wings fore and aft to adjust the position of the centre of pressure relative to the centre of mass. Moving the centre of lift forward or back will cause the bird to pitch up or down respectively. It is effectively modifying its stability as it manoeuvres, rather like when a human leans-forward as they walk and leans-back as they try to stop, moving their centre of mass outside of the contact patch and becoming unstable in the process.

Secondly, birds spread their tails in flight which must contribute to the overall aerodynamic effect but in what way exactly?

They could use their tails passively for stability and actively for pitch control in the same way as aircraft but that would increase drag and waste energy. If the bird tail is used for pitch stability in the same way as it works in aircraft, then the download on the tail would be an extra weight that would need to be supported by the wing, causing extra muscular effort at the shoulders. It may be that there is a trade-off between the effort of moving the wings back and forth for active stability and the effort of supporting the aerodynamic weight of the tail for passive stability. The bird can vary its stability according to what it is doing; whether gently circling in a thermal or actively manoeuvring in combat or display.

4.1 Landing

The tail can also be used to increase lift and/or drag when landing or manoeuvring. However, if the bird is seen to be moving its tail up and down it might not be initiating a pitching motion. The tail could be simply conforming to the relative airflow to reduce drag, while the wing movement achieves the control function. Bear in mind when observing the tail, that the angle of the tailplane relative to the body is not necessarily the same as the angle of attack because the tail is working in the downwash behind the wing.

Different birds use their tails in different ways; albatrosses have small tails with minimum spread and apparently little deflection. Eagles on the other hand have large, widely spread tails which are in constant use, mostly with an asymmetric twisting motion which seems more associated with roll or yaw control rather than with pitch control.

4.2 Tail deflection during hill soaring

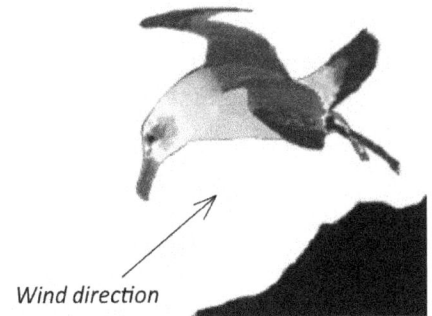

Wind direction

In both birds, while pitching-up for landing, the tail can be seen deflected-down, like an aircraft trailing-edge flap to supplement lift and balance forward swept wings. While hill soaring the tail is upswept to align with the relative airflow and spill lift.

Yaw and roll

Birds don't have ailerons, so they must twist their wings for roll control. Vultures have upturned wing-tip feathers apparently deflected by lift forces, which would assist with directional stability, combined with a dihedral angle for lateral or roll stability. Albatrosses and gulls have swept-back hand-wings for some yaw stability, although they typically have down-turned wing-tips which would tend to reduce yaw and roll stability.

Eagles and vultures have a dihedral angle to their wings and look quite stable whilst circling in thermals albeit with their ever-twitching tail feathers, while albatrosses with drooped outer wings are less stable but more manoeuvrable during dynamic soaring which requires frequent reversals of bank angle.

Despite the complete absence of vertical tail surfaces, birds don't seem to suffer from adverse-yaw or dutch-roll. In fact, they are rarely, if ever, seen to yaw or side-slip; I doubt that anyone has ever seen a bird make a flat, uncoordinated turn or a sideslip during landing. So, do they have a way of controlling themselves in yaw?

Various birds use different techniques and it is difficult to see if they are using their tails to initiate manoeuvres or to trim-out the forces produced by using the wings to manoeuvre. The obvious way to influence yaw is to move the drag-centre by asymmetric movements of head, tail, and feet but it is not clear that they are doing that. The tail could be used as a rudder to yaw the bird in the direction of turn and initiating the secondary roll effect or to correct for adverse yaw during rolling manoeuvres. Raptors are seen to be constantly moving their tails in a twisting manner and that could have both a yaw and a roll effect but the surface area and lever arm of the tail feathers are very small compared with the wings they are trying to control.

Adverse Yaw

One of the difficulties with aircraft control is the adverse yaw associated with roll control. For an aeroplane pilot to eliminate adverse yaw while rolling into the turn, the fix is to apply a small amount of rudder in the same direction as the turn. The Wright brothers found this out the hard way with their first two gliders. They had a wing-warping system for roll control but initially no fin or rudder. Well, if birds do not have fins and rudders why would the Wrights need them? Ironically the Wrights wing-warping mechanism was inspired by watching birds banking into their turns. They reasoned that the birds must be twisting their wings in order to bank and they hit upon the wing-warping bi-plane configuration after playing with a cardboard box in their cycle shop. After experiencing the adverse yaw and lack of control, they had to add a rudder, which cancelled out the adverse yaw and which was operated by the same wing-warping controls. Modern aircraft have better designed ailerons to minimise the problem.

4.3 Wright glider wing-warping

Twisting and Folding

So how do birds get around the adverse yaw problem? There are possibly four ways. Firstly, if wing twisting is used for roll, then the tail feathers could be used as a rudder to cancel adverse yaw. But this seems unlikely because the wing is a relatively large and powerful roll control device whereas the tail is short and relatively weak as a rudder. And anyway, why did natural selection not favour the evolution of a proper vertical-tail, like aircraft have?

Secondly, looking at raptors soaring it is seen that the wings are relatively still and held in a dihedral angle while the tail feathers are constantly twisting the tail surface. In avian flight there is an energy penalty in keeping the wings extended, so a permanent download on the tail will be an undesirable waste of energy. It is more likely that the tail is at a neutral average angle of attack for maximum streamlining and minimum drag.

So how can the tail be used to control roll without causing adverse yaw? It works like this (Figure 4.4). When the tail twists to the right, the asymmetric angle-of-attack induces an overall roll to the left. As the left wing goes down, the angle-of-attack increases on that side creating more lift and drag. The lift acts as a roll-damping force and the increased drag causes yaw in the same direction as the roll. This is similar to what happens when a hang-glider turns. This is good for relatively mild roll rates or small adjustments to bank-angle as for example when thermal soaring. More aggressive manoeuvring would require a more powerful roll mechanism as shown later.

4.4 Tail twist for roll and yaw control

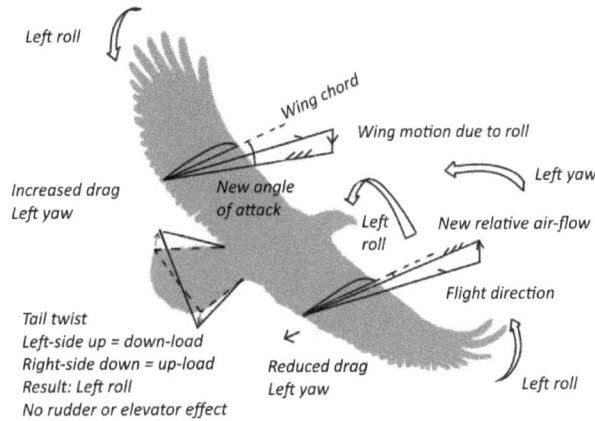

Left roll

Wing chord

Wing motion due to roll

Left yaw

Increased drag
Left yaw

New angle of attack

Left roll

New relative air-flow

Flight direction

Tail twist
Left-side up = down-load
Right-side down = up-load
Result: Left roll
No rudder or elevator effect

Reduced drag
Left yaw

Left roll

Thirdly, bird wings can be twisted but also folded; the folding acts at each joint to change the incidence, the dihedral angle and the sweep angle. For example, to roll left, the wings are twisted left-wing leading-edge down (pronated) and right wing leading edge up (supinated). This decreases the lift on the left wing and increases the lift on the right wing causing the roll to the left. This should cause adverse yaw opposite to the roll. However, if the left wing is extended to increase the semi-span and the right wing is folded to reduce the semi-span, it could move the drag-centre towards the left wing to cancel out the adverse yaw. Compare the picture of the albatross in figure 4.5 and the Wright glider in figure 4.3. Both are in a bank to the right but applying left roll inputs. In both cases the twist of the wings can be seen as the right wing appears thicker and the left wing appears thinner, right leading edge up and left leading edge down. In the case of the Wright glider, it has a vertical rudder which is connected to the wing-warping mechanism to counteract the adverse yaw. The albatross has no rudder so that it may be using the wing folding mechanism to reduce the adverse yaw.

4.5 Neutralising adverse yaw

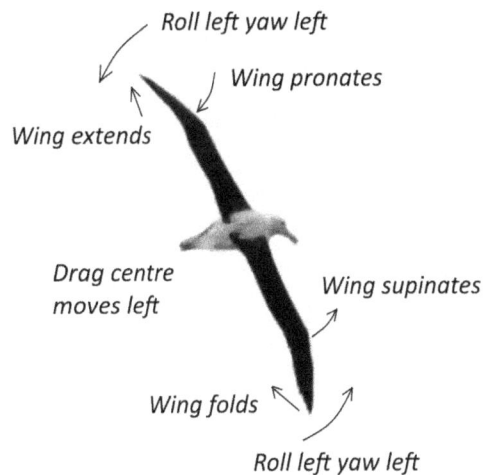

Roll left yaw left

Wing pronates

Wing extends

Drag centre moves left

Wing supinates

Wing folds

Roll left yaw left

This effect can also be seen when birds such as raptors or corvids roll violently in air combat. In figure 4.6, the eagle is flying in a steady glide towards the viewer and is fending-off other birds which are mobbing it. The sequence is read from the top, clockwise, it does not actually lose height – imagine each image superimposed on the previous one. The eagle, looks up to its right. It twists its wings left wing leading-edge down and right wing leading-edge up and begins to roll left. Then it folds and twists the right wing and rolls rapidly to its left, extending its talons. Finally, both wings are partly folded to balance the aero-forces and stop the roll, then extended into the glide posture. This is slightly counter-intuitive because the bird is rolling towards the extended wing with the up-going wing folded. However, this system ensures that the drag of the up-going wing moves inward, towards the bird's centre-line minimising adverse yaw.

4.6 Wing folding for roll control

The wing-folding, twisting geometry that achieves this result is being used here asymmetrically to achieve roll control. The same geometry is exactly what is needed in flapping flight but used symmetrically, to achieve lift and thrust on the down-stroke and lift and minimal drag on the up-stroke. Furthermore, the same folding geometry is used to stow the wings neatly after landing.

Proverse Yaw

Fourthly, it may be that albatrosses deal with this in a different way. Traditionally in aircraft design, the ideal lift distribution was thought to be an elliptical distribution with lift reducing to approximately zero at the tips to minimise induced drag. Positive lift at the wing tips is what causes adverse yaw during aileron deflection. However, if the wing tips are washed-out or pronated to a negative angle of attack giving negative or downwards lift at the wing tip, then roll-control inputs will cause proverse yaw, that is yaw in the same direction as roll.

This has been demonstrated by Nasa with flying models of the swept-back all-wing type without vertical surfaces. With a down-load on the tips, when ailerons are used to roll these aircraft, the increase of down-force on the down-going wing-tip causes an increase of drag on that tip, which is proverse yaw. The opposite effect happens on the up-going wing tip; the download and drag on the up-going wing is reduced, causing roll and yaw in the same direction. See figure 4.7

4.7 Avian proverse yaw

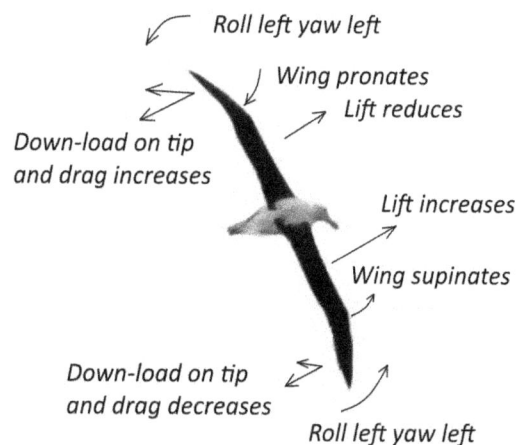

Roll left yaw left
Wing pronates
Lift reduces
Down-load on tip and drag increases
Lift increases
Wing supinates
Down-load on tip and drag decreases
Roll left yaw left

There is a slight form-drag penalty due to the wings carrying the weight of the down-force at the tips. However, this lift distribution will minimise lift-induced drag, while the weight of the download at the wing tips is supported by the lift of the inner wings so that there is no muscular penalty at the shoulder and it might even relieve some of the bending loads there. Albatrosses normally fly with their wing-tips drooped, so this may be what they are doing and, if the energy is free in dynamic soaring, then the drag penalty will not be a problem.

Wing folding

Bird wings are interesting structures because a single basic mechanism achieves several different functions: gliding, control, flapping and stowage. The wing comprises three parts; the upper arm, the lower arm and the hand. When the joints flex, the available angular movement at each joint is divided between swinging fore and aft about a vertical axis, bending up and down about a longitudinal (fore and aft) axis and twisting about a span-wise axis, depending on the axis of rotation at each joint. These complex motions can be replicated in a limited way in a model articulated wing using pin joints. The axes of the joints are not parallel but are skewed relative to each other.

If all of the axes are parallel and the deflections are in the same sense, then the sum total deflection will be large, for only a small deflection at each joint; as it is in incidence. In dihedral, as the wing goes up at the shoulder, the elbow and wrist joints fold down but with a delay, creating the span-wise wave motion to relieve the effort. Where the axes are not parallel and the rotation at each joint is in the opposite sense then the total effect is that the wing folds in a Z formation in plan form with opposite rotation at each joint.

When they are extended symmetrically for gliding flight, wings can be operated symmetrically for pitch control, for varying the wing area or aspect ratio, or they can be operated asymmetrically for roll control. In flapping flight, the wings extend symmetrically and pronate for the down-stroke; then they fold symmetrically and supinate for the upstroke.

Finally, the same geometry allows the wing to fold completely and envelope the body for roosting or promenading on the ground. Thus one set of joints, one mechanism has been modified by natural selection to achieve the functions of lift, thrust, control, stability and insulation not to mention courtship and display. But not necessarily in that order!

Lift distribution

In aircraft design, the wing area used in the lift and drag formulae includes the part of the fuselage between the wing roots, both for high-wing or low-wing aircraft. This is because it is found that the fuselage contributes to lift to some extent and this method provides a good approximation of the effective wing-area. So, it would be fair to say that bird bodies probably contribute to lift in the same way.

The distribution of lift and drag or thrust over the wing will vary according to the phase of flight. In gliding flight, the whole wing produces lift. It would make sense if the lift distribution is biased towards the inboard part of the wing as this would reduce the muscular loads needed to keep the wing extended and also would reduce lift-induced drag which is produced mostly at the tip. Paradoxically, most birds have upturned tip feathers indicating a good bit of lift being produced at the tips and many birds, especially owls, glide with wash-in at the tips meaning that the angle of incidence is greater at the tips than at the wing-roots. Aircraft typically have wash-out at the wing-tips to reduce the angle of attack and improve handling, particularly to prevent wing-drops at the stall. It is possible that seabirds with drooped wing tips may have negative lift at their wingtips. This would relieve bending loads at the shoulder joint and reduce adverse yaw during rolling manoeuvres without reducing the beneficial effects of high aspect-ratio (long, narrow wings).

Terrestrial soaring birds typically have emarginated feathers at the wing tips. The first wing-tip feather is bent the most because it is at a large angle of attack, but working on its own it has a high aspect ratio and produces little drag. Each subsequent feather is working in the downwash of the preceding feather and is working less hard and producing less lift and drag and therefore bending less. Thus the bird gets the benefit of both low wing loading inboard and high-aspect ratio at the tip enabling them to fly slower with a minimum rate of descent to exploit thermals. On the other hand, sea-birds have pointed wing tips. This may be because seabirds fly faster and need a shallow glide angle to exploit dynamic soaring.

Formation flying

Many birds, but not albatrosses, indulge in formation flying in cruising flight as a way to save energy. How does this work? A flying machine will normally produce a small up-lift in the air ahead of the wing and a larger down wash behind the wing. The wing tip vortices repeat this effect with rising air outboard of the wing tips and descending air between the wing tips. Flying in line-astern is definitely to be avoided!

The total effect is that a wave is created which travels with the bird, similar to the V-shaped bow wave of a ship. The influence of this wave spreads out sideways and behind the wing tips. It is this wave that the birds in formation are riding and each bird reinforces the wave for the next bird in the line. (Figure 4.8)

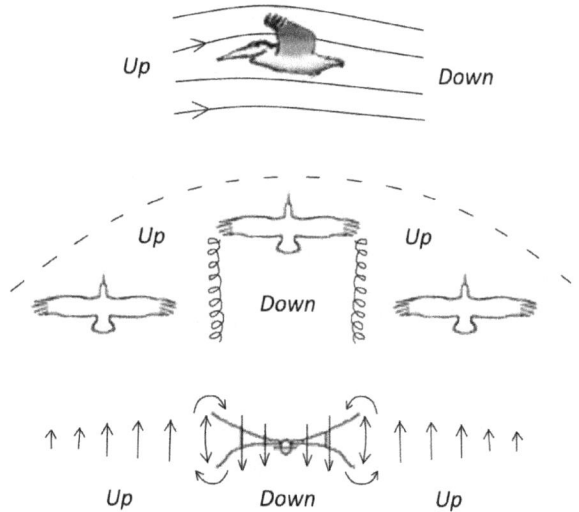

4.8 Formation flying

Chapter 5

Soaring

Hill soaring

In 1901 the Wright brothers began their experiments by flying an unmanned biplane kite. They then progressed to kite-ing on a windward-facing slope of the coastal dunes at Kittyhawk, North Carolina, a location they had chosen after consultation with the United States Weather Service to find a place with strong and consistent winds. Although it was mostly unsaid, they understood that learning to fly was going to be a risky and lengthy business and in order to discover the secrets of flight they would have to survive the inevitable crashes. They chose a windy, hilly and sandy location in order to fly at low level at relatively slow forward speeds while achieving sufficient airflow over the wings to achieve lift and control. They progressed to flying the kite-glider with lines to control the wing-warping and then moved on to manned free-flight in 1902.

After making the first flight in the powered Flyer in 1903, they went home to Dayton, Ohio to develop the powered machine but they occasionally went back to Kittyhawk to practice soaring with the glider. Before the First World War, they achieved the record for soaring flight at about 11 minutes, and held it until gliding resumed after the Armistice.

Aviation development was almost entirely devoted to military powered flight until 1918 when WW1 came to an end with the Treaty of Versailles. Germany was prohibited from building powered aircraft; and sport gliding began to be developed. Gliders were launched with a rubber bungee cord from the tops of hills to glide down to the fields in the valleys. It was soon found that winds blowing up the hill would keep the gliders aloft and it was then possible to fly back and forth along the windward face of the hill. Today, high performance sailplanes are capable of flights of hundreds or even thousands of kilometres along whole mountain ranges.

Birds perform similar soaring manoeuvres. Kestrels are seen to wind-hover, facing into the wind blowing up a hill or embankment; they keep their heads absolutely stationary in order to search for their prey, while the body, wings and tail are constantly in motion. to compensate for the gusts. Finally, the prey is spotted and the bird plunges down to seize the hapless creature and carry it way. Crows soar and tumble in the gusty updrafts on a windward cliff edge with no obvious purpose other than the sheer fun of it. Gulls are typically seen at coastal sites cruising effortlessly along the seawall or sea-front buildings looking for an unguarded take-away.

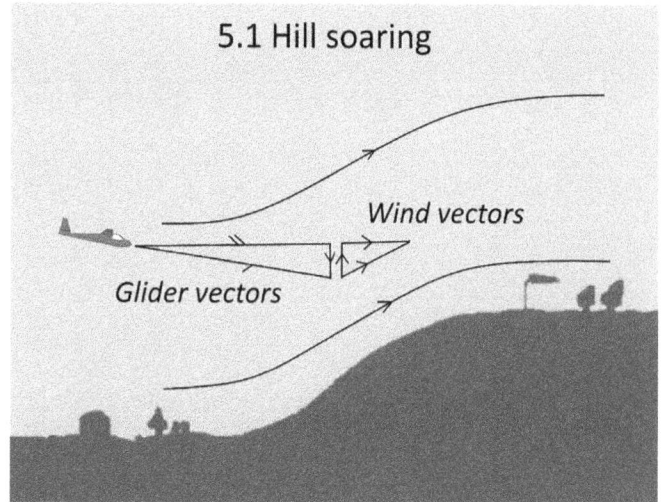

5.1 Hill soaring

Wind vectors

Glider vectors

Hill soaring works because the wind blowing against a hill is deflected upwards and has a vertical component of speed greater than the downward speed of the glider (the sink rate). (Figure 5.1) Although most of the wind is deflected upward by the terrain, not all of the air goes up and over the hill; some is deflected sideways by the contours of the terrain or channelled up gullies. The best hill-lift for sailplanes is found where the wind blows at right-angles to a long, straight hill with relatively smooth contours. Even then, the power of the lift will depend on other factors such as the temperature gradient; which can result in a layered structure in the atmosphere with different degrees of stability or instability. Any pre-existing wave motion in the air may or may-not be in-phase with the hill and may enhance or supress the lift. Enhanced hill lift can take a glider to thousands of feet above the hill and transition into atmospheric wave at much higher levels. Supressed lift can mean that lift is only found close to the hill-face and below the top of the hill. These issues do not seem to bother birds who are able to manoeuvre much more quickly and much closer to the ground.

5.2 Lift/drag Ratio

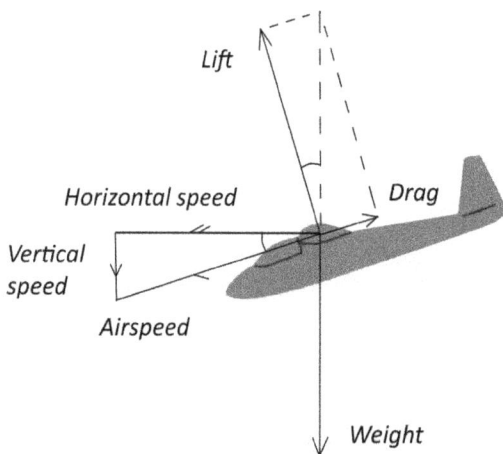

Lift

Horizontal speed

Drag

Vertical speed

Airspeed

Weight

Lift/drag = glide ratio
= horizontal speed / vertical speed

Looking at the diagram of gliding flight (Figure 5.2), the lift and drag forces are seen to be components of the resultant, which is equal and opposite to the weight. The other way of looking at it is that the lift and drag forces are balanced by components of the weight. The glider is in a nose-down attitude because it is effectively going down a slope propelled by gravity at constant speed. The glider is in equilibrium and is not subject to acceleration. The triangle of velocities in a vertical plane shows the air-velocity of the glider angled down relative to the air, together with the vertical component of the glider's airspeed. The third side of the triangle gives the aircraft ground-velocity which is the horizontal component of the air-velocity. But although the glider is clearly going downhill within the air-mass, there is nothing to indicate whether the glider is going up or down relative to the ground.

In figure 5.3, compare a powered aircraft maintaining level flight in still air with a glider maintaining height in an updraft. Both are flying close to their best lift/drag angle of attack, measured relative to the direction of the air-velocity. The powered aircraft is clearly expending energy in the form of thrust balancing drag, to maintain itself in level flight. The glider is also maintaining speed and level flight but is neither gaining nor losing kinetic or potential energy. So is there an energy balance to be accounted-for here? Yes, the glider is expending energy in the form of drag and is propelled by a component of its own weight.

The glider is supported by aerodynamic lift within the rising air-mass so that any unit volume of air containing the glider will weigh more than the same volume of air without the glider. The rising air-mass, having been deflected by the terrain, is doing work against gravity by virtue of its own momentum. That momentum of the wind is caused by the temperature and pressure gradients elsewhere within the atmosphere. The air-mass containing the glider has greater mass than the air-mass without the glider and therefore does more work against gravity and loses more momentum for a given gain of height compared with an air-mass without the glider. The difference in momentum amounts to a difference of wind-speed which equates to a loss of the kinetic energy of the wind downwind of the glider.

**5.3 Level flight
under power and in a glide**

Under power in still air
Air velocity and actual velocity the same

Gliding with a vertical wind
Actual-velocity horizontal

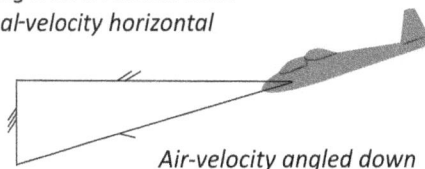

Air-velocity angled down

Now, air density is typically about 1.2 kg per cubic metre, so that a 100m cubic volume of air has a volume of 1,000,000 cubic metres and a mass of about 1,200,000 kg. A glider has a mass of about 400kg, so that the presence of the glider in such a volume of air is not going to make much of a difference and the wind speed is not going to change by much. I mention this because in soaring there is always an energy balance that needs to be accounted for. This is equally important to understand in dynamic soaring. All this happens invisibly and nobody is bothered, except you, dear reader and me.

Mountain Wave soaring

With the development of high altitude flight in the 1940s the motion of the middle atmosphere could be explored and two important effects were discovered.

The jetstream is a giant river of air flowing at high speed along the boundary between the relatively warm temperate zone and the colder polar region. The boundary is where the weather fronts are formed, which cause all of Britain's variable weather.

The other effect is atmospheric wave or mountain wave, which comprises a succession of standing waves downwind of hills or mountains often marked by stationary stacks of lenticular clouds. The standing waves are formed when the wind blows up and over a line of hills or mountains. On the downwind side, the descending air bounces back up and down, forming waves which effectively move upwind at the same

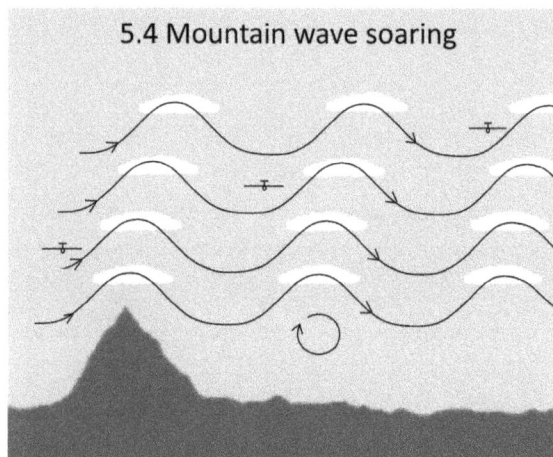

5.4 Mountain wave soaring

speed as the wind and therefore remain stationary relative to the ground.

The conditions required for the formation of mountain waves are: a wind direction which is at right-angles to the line of hills and fairly steady in direction at successive higher levels and wind speed which increases steadily with height. The temperature gradient needs to vary sufficiently to promote layers of stable and unstable air.

Mountain wave is exploited by glider pilots to achieve long distance flights of thousands of kilometres and great heights exceeding 30,000 feet at the top of the troposphere, the part of the atmosphere where most of the weather is generated. Current research with specialised sailplanes is getting up to 60,000 feet in the stratosphere and aiming for 90,000 feet.

Birds have been encountered by aircraft at high altitudes exceeding 20,000 feet and they are known to cross mountain ranges like the Himalayas. Whether they are at such height by desire or accident is not clear. Birds certainly use hill lift which can transition seamlessly into mountain wave but whether they use mountain wave systematically is not certain.

Thermal soaring

Back in the 1920s while hill-soaring, glider-pilots discovered gusts of wind which carried them up and out of the hill-lift. These are thermals: bubbles or columns of air which are warmer and less dense than the surrounding air and therefore buoyant and inclined to rise like a balloon. Thermals are typically marked by cumulus clouds and, given unstable conditions, are prone to develop into towering cumulus or cumulo-nimbus clouds and showers. At last, we can explain the piloting of Peel's pirouetting pelicans! (Figure 5.5).

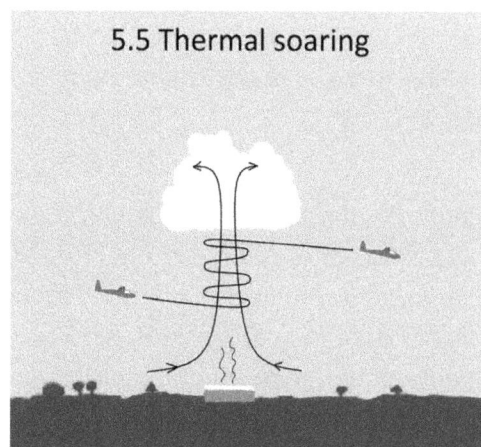

5.5 Thermal soaring

Environmental lapse rate (ELR)

Normally, the ambient temperature in the atmosphere is less at greater heights. The temperature gradient in the static atmosphere is called the environmental lapse rate (ELR). It does not have a fixed value and normally varies with height according to circumstances. If the temperature starts to increase with height this point is known as an inversion.

Dry adiabatic lapse rate (DALR)

Thermals can be triggered by hill-lift or by frontal systems and also by patches of ground warmed by the sun. When the warm bubble of air rises, it begins to expand because atmospheric pressure reduces with height. As it rises and expands adiabatically, that is to say without exchanging heat with the surrounding air, the thermal cools at the rate of about 3 degrees centigrade per 1000 ft. This is the dry adiabatic lapse rate (DALR) which happens when the air is unsaturated with water vapour and no cloud is forming. Depending on how warm the thermal is compared to the surrounding air, the thermal will keep rising until the temperatures and densities equalise, at which height the thermal will stop rising and disperse. (Figure 5.6).

5.6 Moderate ELR
Conditional stability

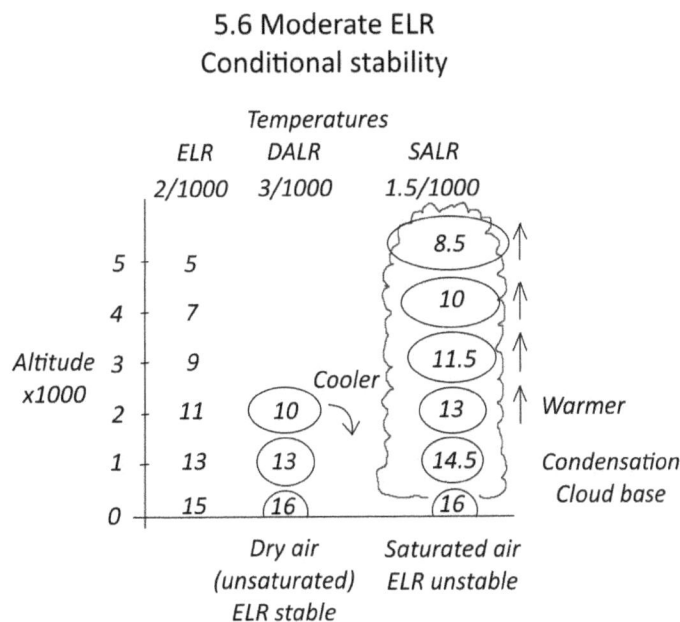

Saturated adiabatic lapse rate (SALR)

Water vapour is a colourless, odourless gas. On the other hand, visible clouds are made of liquid water droplets suspended in the air. The amount of water vapour that can exist in air, before it condenses into liquid water, depends on the temperature; the greater the temperature, the more water vapour the air can contain. When the air contains the maximum amount of water vapour allowed by its temperature, then it is said to be saturated.

If the rising bubble of air cools sufficiently to reach water vapour saturation, then water droplets begin to condense and cloud begins to form; the cloud base marks the condensation level in the atmosphere. As the cloud forms above this level, the condensing water vapour gives up its latent heat of vaporisation and the heat added to the air reduces the rate at which the bubble cools. As it rises, it now cools at the saturated adiabatic lapse rate (SALR) of 1.5 degrees C per 1000ft.

The temperature of the rising air-bubble is initially greater than the surrounding air. If the ELR is less than the SALR of 1.5 degrees C per 1000ft, then the saturated bubble will cool faster than the surrounding air and will eventually be cooler and denser than the surrounding air and will stop rising. This is absolute stability.

An ELR of less than 3 degrees C per 1000ft but greater than 1.5 degrees per 1000 ft is called conditional instability because the instability depends on how moist the air is and how quickly a bubble of that moist air reaches saturation as it cools. In figure 5.6, dry (unsaturated) air is stable while saturated air immediately starts condensing and is unstable.

When the ELR of the air surrounding the thermal is greater than 3 degrees C per 1000 ft, called a steep temperature gradient, the rising air will always be warmer than the surrounding air and will keep rising. This is known as unconditional instability. (Figure 5.7).

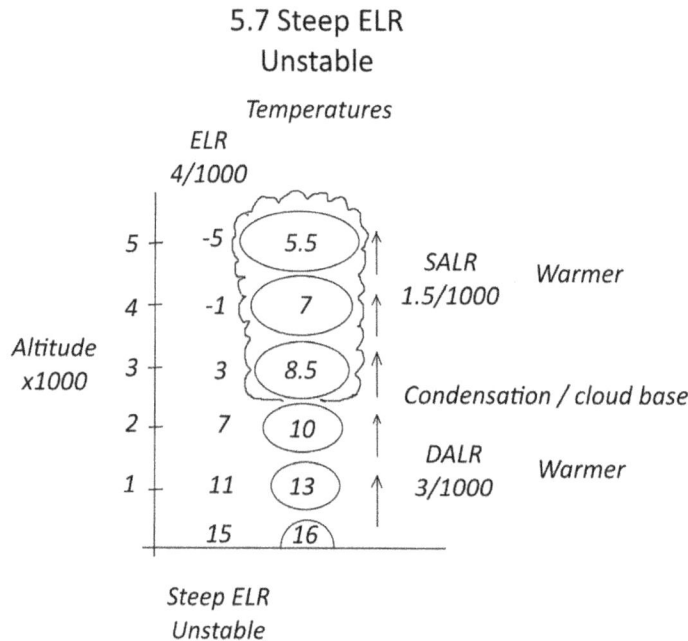

5.7 Steep ELR
Unstable

Here the moist but unsaturated air initially cools at the DALR. As it cools it can hold less water vapour and becomes saturated at the condensation level. Cloud starts to form and it now cools at the SALR but cools less than the surrounding air and so continues to rise. These are the conditions which lead to towering cumulus and cumulo-nimbus clouds.

The standard atmosphere

The ELR of the International Standard Atmosphere, which is a theoretical average atmosphere used to calibrate flight and meteorological instruments, is 1.98 degrees C per 1000ft and this is in the conditional instability range. Generally, the degree of instability, on any particular day or time of day, depends on the ELR and the humidity. The greater the value of the ELR, or the steeper the temperature gradient, the more unstable the atmosphere is, depending on how moist the air is.

Clouds

The type of cloud formed depends on the ELR. When the ELR changes with height forming layers of instability, stratus or layered cloud forms. When there is a deep layer of instability, cumulus or heaped cloud forms with greater vertical extent. The deepest layers of instability lead to cumulo-nimbus clouds. These are rain bearing clouds with thunder, lightning, turbulence, ice and with tops above 20 to 30,000feet.

Soaring birds

Thermal soaring is used by birds for gaining height during hunting, scavenging or migration and it would seem, for fun. Thermals are common over land and thermal soaring is typically associated with terrestrial birds. Vultures and eagles have to gain height in order to spot their prey visually then glide to their target in relatively light winds over relatively short distances. With a glide ratio of say 1 in 13 a raptor could glide about 5 miles from 2000 feet above ground level. The picture is of a snake eagle.

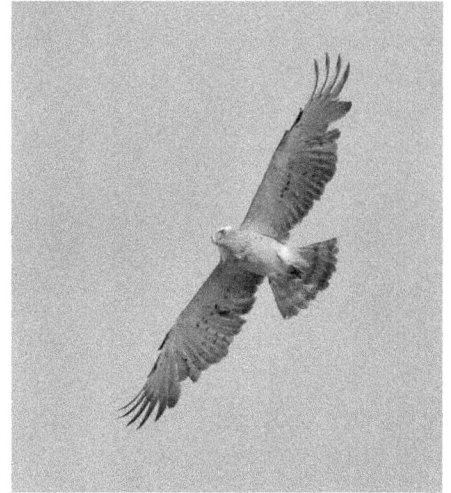

Thermals are less common over the sea because of the more even distribution of heat in the surface waters and the more level surface. Over the sea, clouds and precipitation are more associated with frontal systems. Birds that rely on thermals during migration will typically avoid long sea crossings hence the migratory routes between Europe and Africa will take a Western route via Gibraltar or an Eastern Route via Turkey.

How do they find thermals? Glider pilots find thermals by looking for the cues that lead to thermal production – temperature contrast on the ground and cumulus cloud forming aloft. Where the skies are clear and the ground relatively uniform but hot, thermals are highly likely and can be found at random simply by flying a straight line or by chasing other gliders! Birds typically find thermals by flocking behaviour. They spread out until one bird runs into a thermal and starts to gain height and all the other birds then converge on it.

Soaring is a mixture of passive and active elements. Whereas regular soaring uses the vertical motion of the atmosphere to sustain the glider in flight, it is a passive process in that the aircraft need only be flown in equilibrium and the rising air will do the rest. However, the lift, or rising air, that is exploited in regular soaring does not occur everywhere; it must be actively hunted down by the pilot. Meanwhile, the aircraft must be flown in a way that keeps it within the rising air. In dynamic soaring the wind is everywhere and does not need to be sought. Finding the wind is the passive element but on the other hand, dynamic soaring is an active process in that the glider must perform a specific turning manoeuvre to achieve the desired effect of exchanging momentum and kinetic energy with the wind.

Albatrosses

So far as is known, albatrosses do not do thermal soaring which essentially involves circling in rising air which is typically marked by cumulus cloud. It has been presumed that this is because thermals are less frequent over the sea compared with over the land. In fact, thermals and cumulus cloud do occur randomly over the sea but are typically associated with showers and frontal systems.

Albatrosses, having higher wing-loading and greater flying speeds compared to raptors like eagles and vultures, are less well adapted to circling in thermals but if manned sailplanes can do it then why not albatrosses? It has been found that manned gliders typically need to bank at about 45 degrees to stay in a thermal; at 45 degrees angle of bank the load-factor is 1.41G. This is not a problem for a sailplane, it is just a question of balancing the loss of glide performance with the ability to stay in the core of the thermal and gain the maximum rate of climb. However, for a bird a load factor of 1.4G represents a substantially increased effort to keep the wings extended and a real energy penalty, compared to 1G dynamic soaring. Thermal soaring birds are typically seen to bank at lesser angles; a 30 degree bank has a load-factor of only 1.15G; a relatively low wing-loading and slow airspeeds will give a small radius of turn and keep the bird in the core of the thermal. At albatross airspeeds the turning circle will be much larger and it will be more difficult to stay in a thermal.

The other reason albatrosses do not do thermal soaring is simply that they have a different foraging strategy. This involves travelling much greater distances between isolated feeding sites, which they find by intercepting scent trails drifting downwind close to the surface. Natural selection has favoured the evolution of dynamic soaring in consistent wind-fields over millions of years. The wind-fields have themselves formed over the same time period due to the formation of the oceans because of continental drift. Albatrosses are solitary travellers and it is unlikely that they could quickly adapt to a different foraging strategy if their favoured winds were replaced by thermals.

Ocean wave soaring

Pelicans have been seen flying parallel to the Californian coast, soaring long, straight swells which are approaching the shore, possibly with an off-shore breeze. They do this as individuals and in flocks where the drift angle enables then to fly in echelon, that is a diagonal formation and gain the same advantage as birds flying at height in V-shaped formations. Figure 5.8 shows how the birds lay-off drift in an off-shore wind in order to fly parallel to the wave as it approaches the shore. At the same time, the combination of drift and the echelon formation enables them to avoid flying directly behind the bird in front and thus benefit from the uplift outboard of the wing-tips. As the swell approaches the shore and before it can break, the birds gain height and climb up and over the swell, moving sea-ward to pick up the next incoming wave. In such a manner the birds are able to travel great distances parallel to the shore with little effort.

Waves and swells are caused by the wind and will be mostly moving downwind. A wave moving opposite to the wind will be quickly suppressed. The swells may be caused by weather systems hundreds of miles away and the wind may subsequently change direction; or the waves may travel to a location where the wind is in a different direction. Either way, the wind impinging on a wave which is moving downwind will not have the same relative velocity as that found on the windward side of a stationary hill or a wave moving upwind and will not produce the same vertical motion.

In figure 5.9 can be seen the interaction of waves approaching the shore with an off-shore and an on-shore wind. The wind speed relative to the wave is greater when the wave and wind are moving in opposite directions. When there is an off-shore wind, typically in the cool of the morning, the birds can soar the in-coming wave. When there is an on-shore breeze, when

5.8 Wave soaring

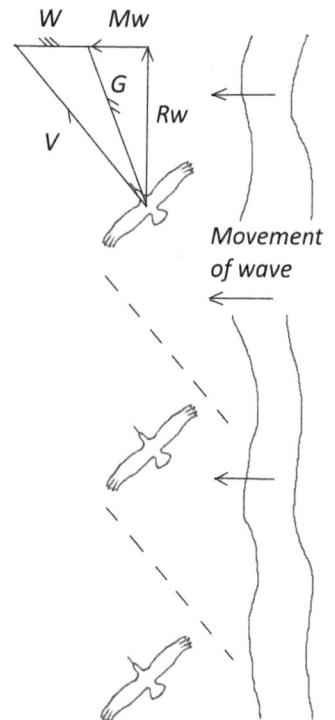

Movement of wave

V   Air-velocity
G   Ground-velocity
W   Wind-velocity
Mw Wave-velocity
Rw Bird-velocity
     relative to wave

the land has heated in the afternoon, there is little lift on the windward side of the wave because the wind and wave are moving in the same direction but instead the birds can soar the coastal cliffs and dunes.

Storm Petrels can be seen in flight at the sea surface, dipping their feet in the water to prevent themselves being blown downwind; effectively acting as a tethered kite.

Albatrosses have been reported as soaring on ocean waves and swells in a similar manner to hill soaring. This is plausible and it is hard to say the observer is wrong; but it is equally possible that they are simply dynamic soaring and encountering the waves or swells at random. They do hill soaring when they return to their nesting sites on land or when they encounter ships at sea. Otherwise, they stay resolutely at surface level and have nothing to do with mountain waves or thermals.

5.9 Relative velocity of wind and wave

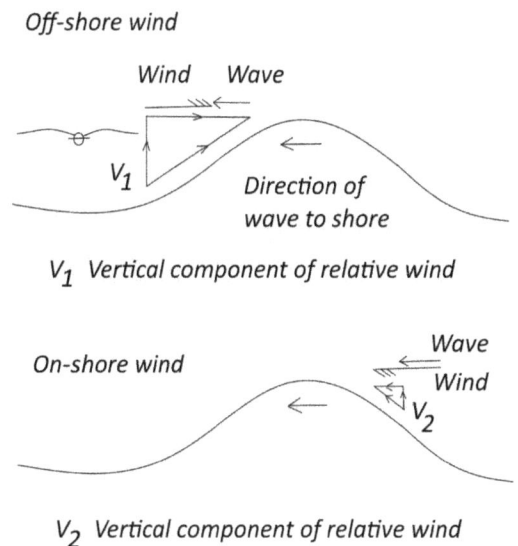

Off-shore wind

Wind   Wave

V_1 Direction of wave to shore

V_1 *Vertical component of relative wind*

On-shore wind

Wave

Wind

V_2

V_2 *Vertical component of relative wind*

If the surface of the sea is going up and down, there will be some vertical motion of the air above the surface but it is not certain that all of the sea motion will be converted into vertical motion of the air – the air could easily be pushed sideways. On land where there are isolated hills, the wind takes the path of least resistance and blows around the hill rather than doing work against gravity by going over the top. This is the case with terrestrial hill soaring which works best when the wind blows normal to a long straight line of hills

Gust soaring

Gust soaring is an idea that appears from time to time in descriptions of bird flight. It is a plausible and attractive theory that perhaps birds can exploit horizontal gusts which boost their airspeed and enable them to convert the excess speed to height.

The problem with this is that, given an average wind speed, gusts will occur equally and unpredictably as both tailwinds and headwinds and cause equal decreases as well as increases of airspeed. However, aerodynamic drag depends on the square of airspeed therefore drag-increases associated with airspeed-increases will exceed drag-reductions caused by similar airspeed-reductions. So, in gusty conditions, average drag will be greater than in calm air and therefore it is difficult to avoid a net energy loss. Nobody ever argued that aircraft fly faster or more efficiently in turbulence!

Dynamic soaring

Dynamic soaring is the main subject for this book and we will return to it in chapter 8. Meanwhile, how does flapping work?

Chapter 6

Flapping

Introduction

There is no doubt that albatrosses spend most of their flight-time gliding rather than flapping. However, they flap when they have to and therefore, it is appropriate to discuss how flapping works. This section is based mainly on my observations and analysis of birds in flight, mostly using slow-motion video replay. Also, I have built some models of articulated wings which replicate some aspects of the motion of bird wings.

Mode 1 and mode 2 flapping motion

There are different modes of flapping which, in other publications, are confusingly known as fast-gait and slow-gait. I will refer to them as mode 1 and mode 2. Mode 1 refers to large, heavy birds flying relatively fast but with relatively slow flapping rates. This is what is illustrated in figure 6.1 and 6.2. Mode 1 enables the wing to produce lift during both the down-stroke and the up-stroke; with thrust during the down-stroke and minimal drag during the upstroke.

Mode 2 is typically performed by small birds with rapid flapping rates, for example, a small song-bird (passerine) in horizontal flight. (Figure 6.3 and 6.4). In mode 2, the wing is extended during the down-stroke, producing lift and thrust but during the up-stroke, the wing is almost completely folded producing no thrust and virtually no lift. This flapping mode will typically be used during a flapping phase comprising several up and down strokes, alternating flapping phases with ballistic or gliding phases, known as flap-bounding or flap-gliding.

Mode 1 flapping motion

It might be thought that flapping is a purely oscillatory motion and would result in a large energy loss due to acceleration of the limb at the end of each stroke. In fact, wing flapping is more like a translating rotation. If you look at a slow-motion video of a large bird in flight and concentrate on the wing-tip, you will see the wing-tip move forward and downward then backward and upward competing a quasi-elliptical locus relative to the body, the circular motion at each tip rotating in the sense of a rolling wheel. This motion through the air, combined with the forward flight of the bird, results in a wave-like motion at the wing-tip. (Figure 6.1).

6.1 Mode 1 Flapping motion

Comparing Figure 6.1 and Figure 3.24, you can see the rotation of the wing-tip is opposite to the aerodynamic circulation. (sorry, the two diagrams are travelling in opposite directions). Does this mean that the circulation is enhanced and the wing works at an artificially high lift coefficient? It is difficult to imagine that natural selection would preserve an unfavourable aerodynamic system; so I am inclined to think that there is some advantage to flapping the wings in this way. It is something worthy of further research.

Now, if you are wondering if this makes a difference to the effort of flapping, try this experiment. Stand with your arm out-stretched to the side and flap it up and down, noting the amplitude of the motion, the effort required and the quizzical looks of your neighbours! Now try swinging your arm around in a rotary motion, achieving the same diameter circle as the amplitude of the flapping and the same frequency. Note the effort required and, by the way, don't knock anything over! You should find the rotary motion was easier. But there is more to it than that.

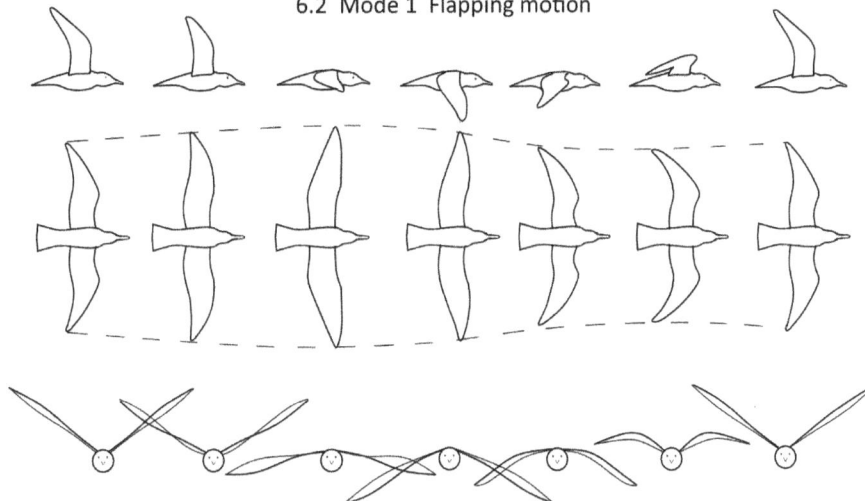
6.2 Mode 1 Flapping motion

Look at the same bird in flight in figure 6.2. Not only is the wing flapping up and down but also it is folding and twisting. The folding reduces the wing span during the up-stroke by flexing the shoulder, elbow and wrist joints. The angular deflection at each joint is divided between incidence, sweep and dihedral axes, resulting in a conspicuous change of dihedral and sweep, with a subtle change of incidence. The angular motion in sweep is in the same direction at each of the shoulder and wrist joints but opposite at the elbow; in other words, when the upper arm (humerus) and hand swing back at the shoulder and wrist respectively, the lower arm (radius and ulna) swings forward at the elbow.

During the down-stroke the wing is fully extended and pronated. The wash-out at the wing-tip (pronation) needs to be greater than at the wing-root because of the greater speed of the wing-tip. This cannot be achieved simply by twisting the shoulder joint. The angular deflection in incidence (angular twist about a span-wise axis) is in the same direction at each of the three joints; the small change of incidence at each of the shoulder, elbow and wrist joints amplifies the overall wash-out from shoulder to wing-tip.

During the up-stroke, the sweep-back angle increases at the shoulder, decreases (sweeps forward) at the elbow and sweeps back at the wrist. This keeps the centre of pressure in approximately the same place while allowing movement of it to control pitch attitude. At the wrist, the increase of incidence together with downward and rearward flexure of the hand-wing, combines to increase the angle of incidence and amplifies the wash-in (supination) of the primary feathers in order to maintain a positive angle of attack during the up-stroke.

Not only does the wing-tip follow a wave-like motion but also the wing itself has a span-wise wave-motion from root to tip. This is because the folding action and the flapping action are not in phase; the maximum and minimum extension occurs during the middle part of the flapping down- and up-strokes respectively. The change of dihedral angle at the wrist is slightly out of phase with the change of dihedral at the shoulder, creating that span-wise wave motion and so relieving the muscular effort at any particular moment and spreading the effort over a longer period of the flapping cycle.

Mode 2

Small birds, despite their apparently frantic flight regime, are not flying as fast as large birds but their small mass means that they do not have very much momentum to reduce their loss of speed between power strokes. With short wings, they have to achieve very rapid flapping rates to enable the wing to produce enough tip-speed and thrust to maintain their average bird-speed. However, a rapid flapping rate means that a large supination of the wing would be needed during the up-stroke to maintain a positive angle of attack and positive lift; otherwise, the wing would produce a down- force. For these birds, the negative effects of the up-stroke are minimised by almost completely folding the wings during the upstroke, relying on body lift and a degree of ballistic flight between down-strokes. (Figures 6.3 and 6.4)

6.3 Mode 2 Flapping motion

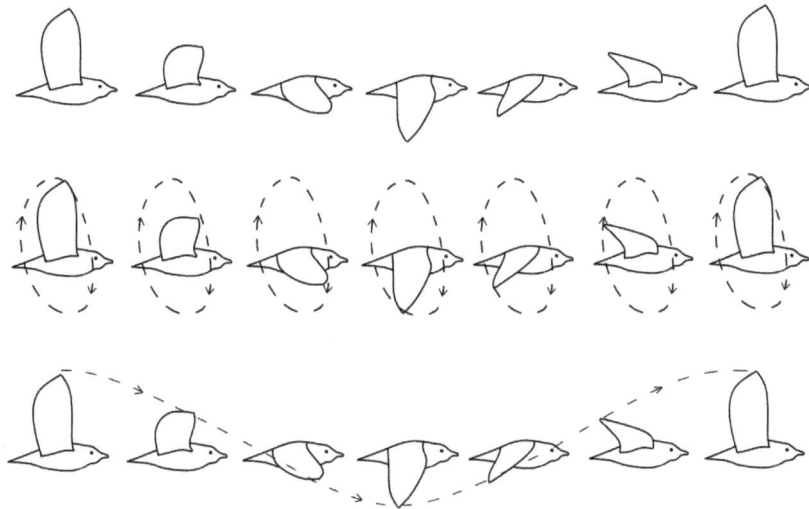

This flapping mode is typically combined with flap-bounding and flap-gliding, in which a few seconds of flapping are alternated with a few seconds of either ballistic flight with the wings folded or gliding flight with the wings extended. In both cases the flapping phase includes a slight pitch-up and height gain. The bounding or gliding phase then results in a slight pitch-down and a loss of speed or height.

6.4 Mode 2 Flapping motion

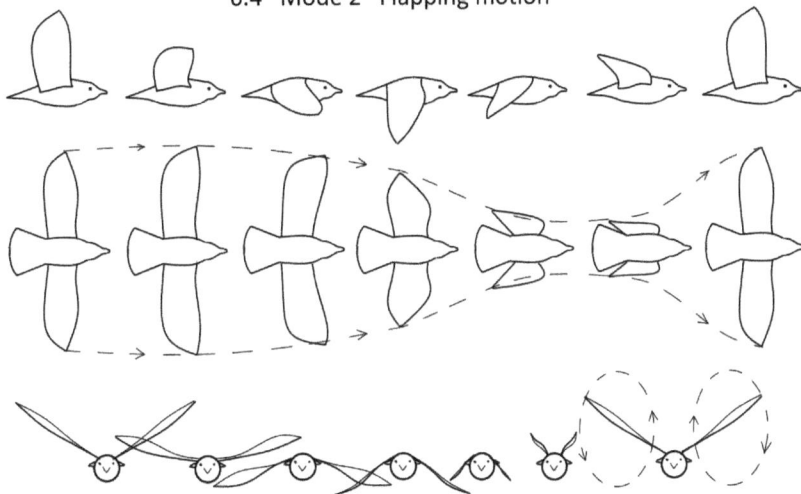

Angle of Incidence and Angle of Attack

The angle of incidence is the angle between the wing chord and some level reference line on the body. In aircraft, this angle is fixed during the construction process. In birds the angle is variable as part of the flapping, gliding or control processes.

Angle of attack is the angle between the wing chord and the relative airflow. In gliding flight, the direction of the relative airflow is opposite to the direction of motion of the body and the wing. In flapping flight, the angle of attack of the wing depends on the angle of incidence and on the vertical motion of the wing and on the forward direction and speed of the bird. The vertical motion of the wing and the angle of incidence will also vary at different stations along the wing-span from root to tip. At any given flapping rate, the wing tips will move faster than the inboard sections of the wing.

Force components during flapping

In flapping flight, the wings must not only be held extended but also flapped, folded and twisted in a rhythmical manner to produce lift and thrust. In flapping, as in gliding, the wing produces a resultant force comprising the lift and drag components which are orthogonal to the airflow relative to the wing. But this airflow is different to the airflow experienced by the body. In other words, the wing is moving in a different direction to the body; the body moves horizontally but the wings move up, down, forward and back.

The wing can produce thrust and drag at the same time. To resolve this paradox, we have to introduce more new terms. Firstly, the lift and drag forces orthogonal to the flapping wing's relative airflow will be called wing-lift and wing-drag. These combine to produce the resultant force. Secondly, the lift, drag and thrust forces orthogonal to the whole bird's relative airflow, produced by the same resultant will be called bird-lift, bird-drag and bird-thrust. (Note that in most of the vector diagrams the drag force is exaggerated for clarity giving an apparent lift/drag ratio of about 4:1. In reality an albatross has an L/D of about 20:1, including the drag of the body, while the wing on its own has a better L/D than that).

The down-stroke in mode 1 and mode 2

For all birds in both modes 1 and 2, the down-stroke in level flight is approximately the same. During the down-stroke, the wing is fully extended, at a small positive angle of incidence at the shoulder, increasingly pronated to a larger negative angle of incidence (wash-out) at the wing-tip. Figure 6.5 shows a typical mid semi-span wing station at a negative angle of incidence but a positive angle of attack.

The wing has a large positive angle of attack, achieved by a combination of wing twisting and the vertical and horizontal motion of the wing and therefore works at a large coefficient of lift. The wing produces wing-lift and wing-drag orthogonal to the relative airflow, combined to give the resultant, which is tilted forward relative to the direction of motion of the body.

The forward tilt of the resultant force ensures that the horizontal component, bird-thrust, goes into propelling the bird forward. In order to maintain average forward speed, the down-stroke must accelerate the bird forward, gaining

6.5 Downstroke

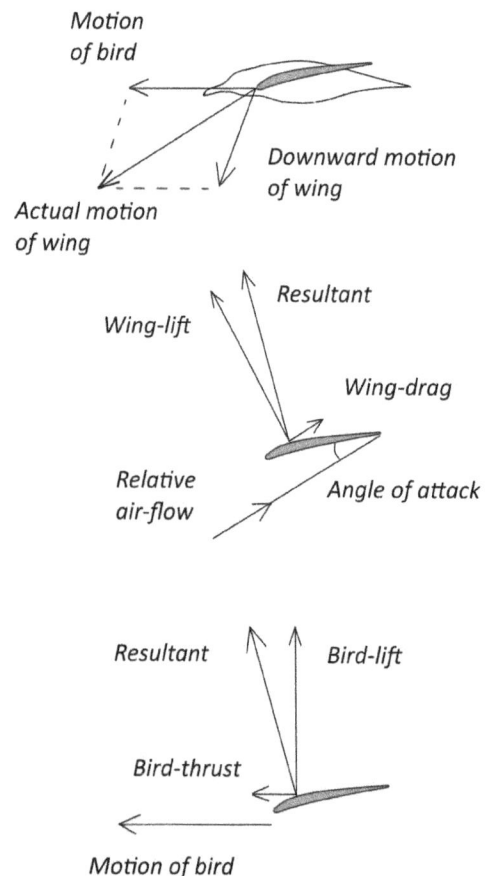

Motion of bird

Downward motion of wing

Actual motion of wing

Wing-lift

Resultant

Wing-drag

Relative air-flow

Angle of attack

Resultant

Bird-lift

Bird-thrust

Motion of bird

speed, to balance the loss of speed due to bird-drag during the up-stroke. At the same time the vertical component, the bird-lift, is limited to that needed to support the weight of the bird to minimise any heaving or up and down motion.

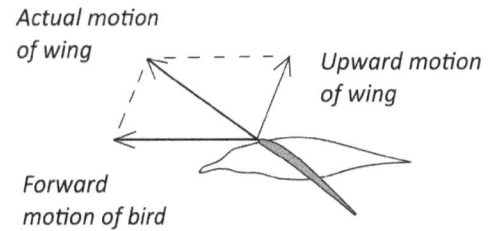

Mode 1 up-stroke

The up-stroke is different for most birds, depending mostly on their size. Figure 6.6 shows again an intermediate wing station, this time with a large angle of incidence but with a small angle of attack. Large birds such as geese, eagles and albatrosses flap relatively slowly and use relatively little folding and twisting in the upstroke, ensuring that the wings produce bird-lift during both the up-stroke and down-stroke. They produce bird-thrust on the down-stroke at the cost of a little bird-drag in the up-stroke.

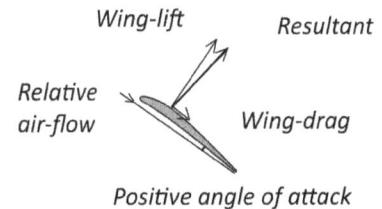

During the mode 1 up-stroke, the wing is folded and supinated (twisted leading-edge up) to allow it to rise at a large angle of incidence but a small angle of attack. This produces bird-lift approximately equal to weight and a small amount of bird-drag because the resultant is tilted backward. This in turn, means that there is a loss of bird-speed during the up-stroke which must be compensated-for by an increase in bird-speed during the down-stroke.

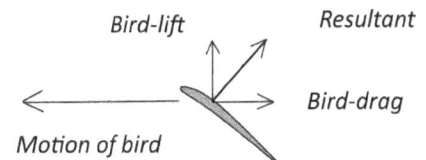

Span-wise force distribution

The inboard section of the wing is subject to relatively little vertical movement and produces a more consistent bird-lift due to the forward motion of the bird. During the down-stroke the relatively little wing pronation in-board means that the resultant produces bird-lift and a little bird-drag but almost no thrust. During the up-stroke, there is bird-lift but some bird-drag is inevitable.

While the whole wing is subject to the forward motion of the bird, the outboard section produces the largest part of the force, due to the greater airspeed caused by the downward and forward motion of the flapping stroke; aerodynamic force is of course, proportional to airspeed squared. The outboard wing is twisted leading-edge down (pronated) in the down-stroke and must work at a large angle of attack to produce the maximum lift coefficient but not so great as to stall the wing. The wing moves forward during the down-stroke which increases the relative airspeed. The resultant is tilted forward producing a large bird-thrust component and keeping the total bird-lift or up-force approximately equal to weight. The bird accelerates forward.

During the up-stroke, the outboard section folds back and down and twists leading-edge up (supinates), while the backward and upward motion reduces the relative airspeed. The outboard section is now working at a large angle of incidence relative to the body but at a relatively small angle of attack to minimise drag. Consequently, and inevitably, the resultant is tilted back but drastically reduced which reduces the bird-drag but also the bird-lift. During the upstroke bird-lift is maintained by the inboard section. Clearly, these effects vary smoothly from wing-root to wing-tip but it is fair to say that most of the bird-lift is produced inboard and most of the bird-thrust is produced outboard.

Power and recovery

The overall motion of the bird, during flapping, is a cyclical fore and aft positive and negative acceleration, rather like the motion of a rowing boat where there are alternating power strokes and recovery strokes. While this is going on, during both the up- and down-strokes, the up-forces are maintained approximately equal to body weight so that the body moves forward smoothly with minimal up and down heaving. The slight heaving motion can be seen during flapping by long-necked birds like swans.

The heavier the bird, the more will inertia resist any change of speed and smooth out the cyclical motion. The smaller the bird, the more rapid is the wing beat frequency to minimise the effect of the thrust and drag pulses.

Internal forces

Aircraft wings are fixed in position more or less rigidly; the forces acting on the wing cause internal stresses and small deflections in the structure of the wing. Effectively, the wing works like a very stiff spring but no energy is expended holding the wing in place and the spring energy stored in the wing structure during flight is released when the flight loads are removed after landing. In aircraft, energy is expended as thrust purely to overcome the drag forces, not to equal the lift forces.

On the other hand, a flying animal expends a certain amount of energy just keeping its wings fully extended, by applying muscular forces at the various joints against the bending loads caused by the aerodynamic forces.

The bird-lift and bird-thrust or the wing-lift and wing-drag forces effectively form components of a single resultant force acting about an intermediate axis, which then produces a bending moment at the shoulder joint. The bird's muscular effort is therefore directed to both move the wing in the desired direction but at the same time to resist the aerodynamic force which is trying to move the wing in a different direction.

Twisting forces

Rigid, cambered wing sections used in aircraft, as well as producing lift and drag, also produce twisting forces about the span-wise axis. Aircraft wings therefore, have to be particularly stiff in torsion to resist this span-wise twisting, especially when the aileron-loads are also applied. An aircraft in a dive has a limiting maximum speed known as the V_{ne}. This structural limit is not due to the amount of lift being produced but rather is due to the twisting forces on the wing and the strength of the tail needed to resist the consequent pitching moments. Aircraft wings are normally straight; and the twisting forces have to be constrained by the internal structure or external bracing.

Birds have the same problem but they have to deal with it in a different way, using a flexible jointed structure. When a bird dives and gains speed, the wings are partially folded. This has several effects:

the wing area is reduced, the span is reduced and the lift distribution is modified fore and aft, apparently using the distribution of the aerodynamic forces to compensate for the increasing twisting forces. Thus the drag is reduced to allow the speed to increase and the bending and twisting loads at the shoulder are relieved.

Bird-wings are substantially made of feathers which are both stiff and to a degree flexible and which are controlled individually by small muscles around their insertion points. If the feathers are able to conform to the airflow by flexing and the bird-wing can be moved fore and aft, this may give bird-wings a neutral pitching moment, leaving the bird able to control the wing twist with minimum muscular effort.

Vortex wake

The consequence of lift-induced drag is that a vortex is shed by each wing-tip. In mode 1, lift is produced continuously in both up and down strokes and therefore, two quasi-parallel, undulating vortices are shed. (Figure 6.7).

6.7   Mode 1   Wing-tip vortices

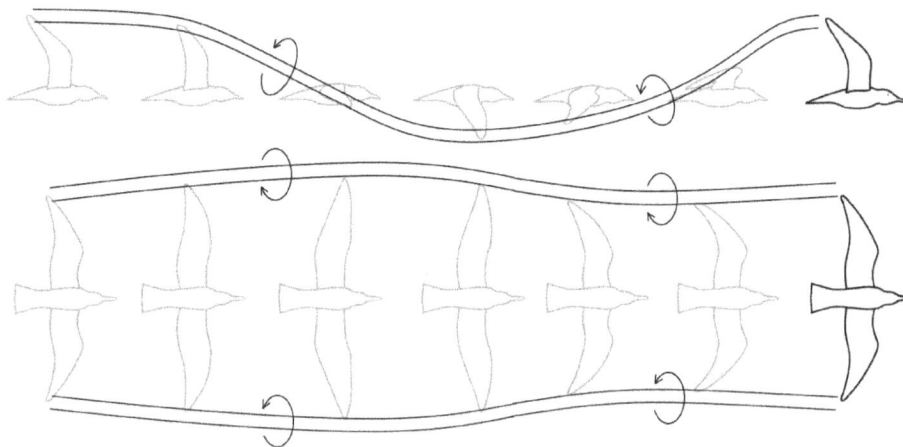

In mode 2, vortices are shed during the down-stroke but not during the up-stroke, because the wings are folded and not producing lift. The vortex trail is thus a pair of inclined arcs; but nature does not allow a vortex to end in a fluid and so the ends of the arcs spontaneously join-up at the top and bottom to form a vortex ring. (Figure 6.8). Notice that the locus of the wing tip is opposite to the rotation of the shed vortex; whereas before we saw that the translating rotation of the wing tip was opposite to the bound vortex. Does this mean that the bound vortex is enhanced and the shed vortex suppressed? Again, it is hard to imagine that natural selection would favour the opposite effect. Maybe it just means that there is no effect? Scientists have carried-out various experiments, visualising these vortex wakes with lasers, flash-guns, particles and bubbles etc, in order to calculate the momentum induced in the air and thus the forces produced by the wings.

6.8   Mode 2   Vortex ring

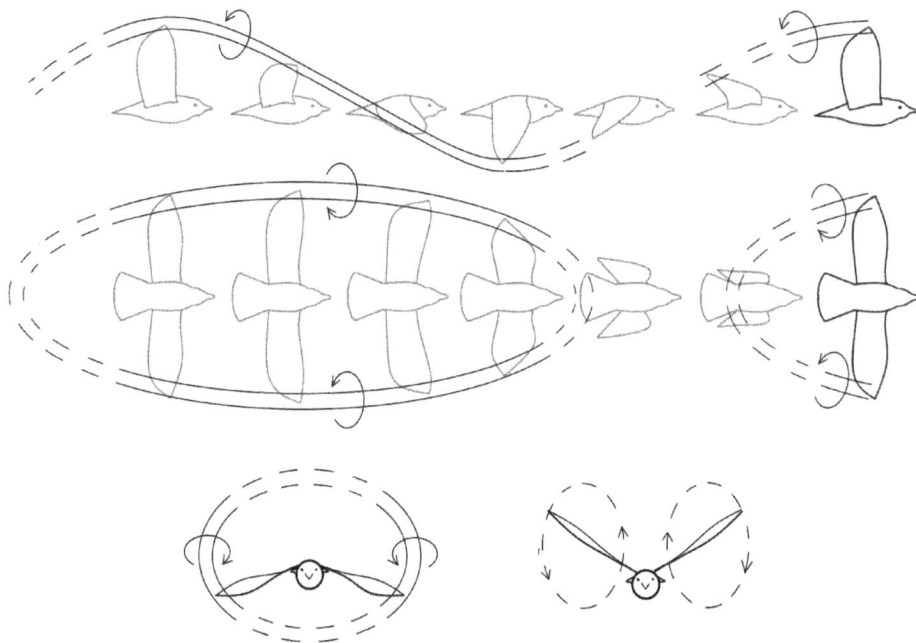

Rotation of vortex is opposite to rotation of locus of wing-tip

Take-off

Take-off involves achieving flying speed, establishing a rate of climb and is normally undertaken facing the wind to reduce the ground-run. Aircraft normally rely on thrust to achieve the necessary acceleration and a long runway to achieve the necessary speed (unless a catapult is being used). During take-off, airspeed is the sum of the groundspeed achieved plus the headwind experienced; so for a given airspeed, the stronger the headwind the less groundspeed is needed.

Most birds use more than wing-thrust for take-off. From a standing start on a perch, most birds will use a combination of leg thrust and a dive to use gravity to assist with acceleration. Taking-off from a surface, a jump, a run or some paddling achieves the necessary speed and/or height from which flapping can take-over.

Pigeons are normally able to launch vertically by using a squat and spring, the wings being unfurled and raised during the squat and powered down during the spring. The body is tilted upward and the wing beat is approximately horizontal. The bird motion is vertically upward and the relative airflow is vertically downward. (Figure 6.9).

6.9 Pigeon take-off

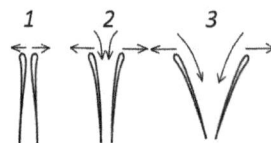

Wing-tips peel apart leading-edge first inducing air-flow into the gap

The clap-fling form of wing motion is used, by clapping their wing-tips together at the top of the up-stroke and at the bottom of the down-stroke. As the wing-tips are peeled apart starting at the leading edge, the pressure reduction between the wing surfaces draws air into the gap and promotes the establishment of circulation around the wings which otherwise might take a fraction of a second longer. During vertical take-off, the wing produces vertical thrust on both the up- and down- strokes, the horizontal bird-up-lift and down-lift then approximately cancels out.

Raptors such as ospreys and sea-eagles are able to launch from the surface of the water by a similar method even with the extra weight of a fish in their talons. After a few strokes with inverted tips during the upstroke, they transition to a more conventional mode 1 and then give themselves a good shake to get rid of excess water from their plumage.

Sea-birds and water-fowl typically use a running or paddling take-off using webbed feet together to give themselves an initial thrust and then paddling with a running motion. (Figure 6.10). The wing motion is modified due to the proximity to the surface The end of the down-stroke is modified with an exaggerated forward sweep of the outer wings; the up-stroke begins with the tips almost inverted to provide thrust but gradually transitions to the regular mode 1 once airborne.

6.10   Flapping take-off

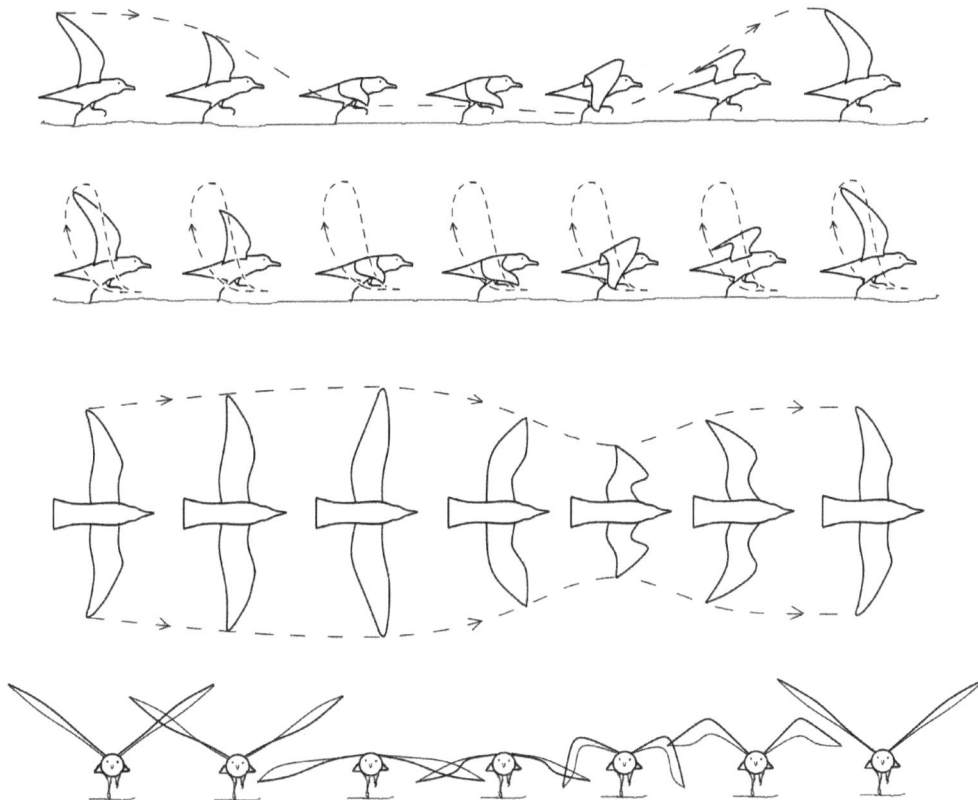

Albatrosses will never be found in the branches of a tree. They typically nest on wind-swept hillsides above coastal beaches and, given a sufficient on-shore breeze, take-off is a simple case of spreading the wings and stepping forward into the rising air flowing up the slope. At some roosting sites there is a sloping open area within walking distance which the birds habitually use as a runway in light winds. In this case a relatively long but down-hill, flapping take-off run is required to achieve flying speed.

6.11 Albatross take-off

During flapping, the thrust depends on the relative airflow over the wings which is a combination of the bird's forward speed and the flapping action. Therefore, the efficiency of the flapping stroke is limited at low airspeeds. Albatrosses have long wings which give them a good glide ratio but have relatively short legs which limit the amplitude of the flapping stroke at the very time during the take-off when maximum thrust is needed to gain flying speed. The lower part of the flapping stroke is modified with a forward instead of downward motion.

You would think that albatrosses will wait-out a calm, floating on the surface rather than waste energy by flapping. The GPS data in chapter 9 suggests that even in light winds they keep flying using a mixture of flapping and dynamic soaring unless there is a purpose in alighting. When necessary, they can launch from a rising wave and quickly get into dynamic soaring. If they have to take-off as an escape manoeuvre in light winds or a calm, then a long take-off run is needed with paddling to assist wing thrust.

Flapping to gain height

As explained in chapter 3, when an aircraft has to gain height, it does not involve increasing lift. Rather, it is necessary to pitch-up and increase thrust. For a bird to gain height, it must pitch-up slightly and increase bird-thrust by flapping harder with increased flapping frequency or increased flapping amplitude. This means that the resultant force is greater and therefore both wing-lift and wing-drag must increase so that the propulsive effect of the bird-thrust is increased while the bird-lift, normal to the upward inclined flight path, is slightly reduced.

Descending

Do birds ever flap when they are descending? If birds in normal flight have a very small range of speeds, then stopping flapping would presumably be the first thing they will do if they want to lose height. In a glide, the wing shape does not correspond to any particular point in the flapping cycle which is characterised by a large span-wise variation of incidence. Instead the wing must be fully extended or partly folded and the angle of incidence and hence the angle of attack must be approximately the same all along the span. The angle of descent can be increased, without gaining speed, by increasing drag, by increasing the angle of attack, by pitching up at reduced airspeed, by folding the wings, by extending the legs and/or by spreading the tail.

Landing

For landing, a suitable spot must be identified and an approach made for touchdown into the wind to minimise groundspeed. Manned aircraft normally make a long, straight and fairly shallow approach to a runway and then a long touchdown and ground roll.

Landing more like helicopters, birds flare to a hover or very nearly zero forward speed and drop onto their outstretched feet from only a few centimetres of height. Birds are of course powered flying machines and can always add a few wing-beats to fine-tune the manoeuvre.

Landing on a perch or a ledge, the bird will normally aim below the spot at relatively high speed and then pitch up, converting speed to height to arrive with zero forward speed. For touchdown on the level, a shorter, slower final zoom brings the bird to the stall at zero forward speed.

A water landing can involve a steep descent to the surface followed by a short zoom, drop and splash-down at zero forward speed. For heavier birds like swans or pelicans a hydro-planing water ski touchdown on extended upturned webbed feet is made. For small birds such as the loon, which has a small wing and therefore a high wing-loading, a high-speed final approach is made, with legs trailing, and a planing touchdown on the chest.

Landing is not the best time of day for an albatross as they are often seen rather inelegantly belly-flopping onto the ground, albeit without any damage. (Or maybe natural-history film makers just like showing the bloopers!) In any event more often than not, the landing ground is tussock grass or sand which provides a relatively soft surface. Landing on water is a relatively straight-forward case of reducing speed, pitching up and extending the landing gear to act as water skis before settling onto the surface.

Albatrosses returning to the roost can use hill soaring to clear the coastal sand dunes and gain height to reach the landing site but then they have some additional problems. Landing, like the take-off should ideally be made into the wind so the groundspeed is the airspeed minus the headwind but that may mean landing downhill. That means that there may be limited wing-tip clearance to make the final turn into wind and no way to use the final swoop to shed the last bit of speed. Flapping at ground level is again difficult due to the long wings and limited wing-tip clearance.

Fluttering

During the landing phase, many birds have a technique which involves fluttering the wings about a span-wise axis at a high average angle of attack, while holding the wing-tips well clear of the ground. The angle of attack appears to be greater than the normal stalling angle of attack and the forward speed is very slow. (Figure 6.12). Wing-fluttering may be a way of rhythmically re-attaching a detached, stalled airflow to achieve an exceptionally high lift coefficient at high angles of attack. Or it may be a way of shedding vortices from the trailing edge to achieve a kind of vectored thrust.

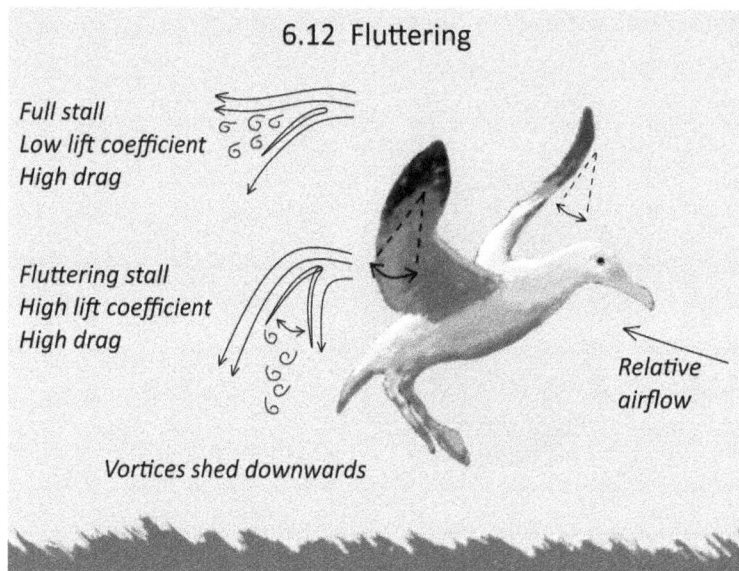

6.12 Fluttering

Full stall
Low lift coefficient
High drag

Fluttering stall
High lift coefficient
High drag

Relative airflow

Vortices shed downwards

Hovering

At the scale of an insect or a humming bird, a very fast flapping rate is possible because of the very small mass of the animal. A fast flapping rate means that the relative airflow is almost entirely due to the motion of the wing rather than the motion of the creature as a whole. The flapping of these small bird's wings provides thrust on both down-stroke and up-stroke, the wings being fully extended during both strokes with the wing inverted during the up-stroke. These strokes being more or less horizontal and the induced, relative airflow more or less vertical. Normal to the airflow, the horizontal forces are minimal, the effects of which readily cancel-out. They effectively fly with vectored thrust. A hovering humming bird produces thrust equal to its weight, augmented by translational lift due to any forward motion. Interestingly penguins and hummingbirds have similar wing actions, producing thrust on both up and down strokes - but there the similarity ends! (Figure 6.13).

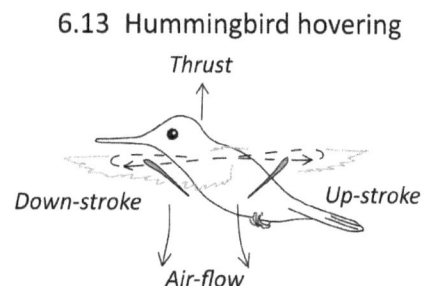

6.13 Hummingbird hovering

Thrust

Down-stroke

Up-stroke

Air-flow

Translational lift is a feature of helicopter flight. When hovering, the rotor system produces essentially downward thrust equal to the weight of the helicopter. In forward flight however, the whole rotor disc and its airflow act as a wing producing a downwash of the relative airflow which supplements the rotor downwash and relieves the power requirement in forward flight.

Stowage

The same folding geometry which is used symmetrically for flapping and asymmetrically for control is also used to completely fold and stow the wings alongside the body, converting the wings into an effective thermal blanket. Can you imagine yourself in an airliner on a night flight and asking the cabin crew for a blanket, only to be told, 'No, the blankets are being used for the wings'!

The way the bird folds its wings is just as difficult to see as the flapping motion. Suffice to say that the three parts of the wing end up as a closely packed bundle and presumably the muscles and tendons which help to hold the wings extended in flight become more or less relaxed when the upper arm is folded back alongside the body.

When I was developing an ornithopter model with a bird-like articulated wing, I built the wing in a fully extended posture with the pin hinges between the three spars. The axes of the hinge pins were not parallel but were variously skewed to give the folding motion required for flapping. I found that the three spars on their own, connected with elbow and wrist joints, easily folded from an extended Z-formation to a tight Z-shape which had a curved and not flat cross-section; exactly what is needed to stow the wings around a bird's body. This led me to think about the evolution of wings. For what was a wing used before it became a flying surface? It could be that wings initially evolved as thermal blankets neatly stowed around the body or extended perhaps for brooding eggs or chicks or for display; they would then be pre-programmed with the geometry necessary for controlled and powered flight, waiting only for sufficient surface area and a suitable centre of gravity.

It is interesting that the Wright brothers developed their aircraft the same way that birds evolved. They built their aircraft from a skeleton of ribs and spars covered with linen sheets. They learned to fly it as a kite and then as a glider before committing themselves to powered flight. So in the evolution of bird flight, it would make sense if the structure and geometry came first; then the physiology and neurology of flight control, while the proto-birds were parachutists and then gliders, then finally powered fliers.

Chapter 7

Navigation

In pilot navigation there are basically two techniques: map-reading and dead-reckoning. Dead-reckoning is the use of plotting charts, the forecast winds and the triangle of velocities to calculate headings to steer, groundspeeds and timing. The triangle of velocity is fundamental to understanding how dynamic soaring works because of the effect of the angle of drift dividing the acceleration vectors; so it will be worthwhile going over the basics.

The Triangle of Velocities

Imagine you want to cross a river in a boat. See figure 7.1. If there is no current flowing, then you simply aim the boat at a point on the opposite bank and set off. However, if the river is flowing and you blindly set off in the same direction, you will get to the other side some distance downstream of your intended target. The difference between the boat's heading and the path it follows is the angle of drift. To make the boat travel direct across the river while the current is flowing, you have to point the boat diagonally upstream using the same angle of drift. You now find that the journey time is longer because the boat travels further through the water compared to the direct path between the two points on the river bank. In other words, the journey time is longer because the actual speed between the two points on the river bank is slower even though the speed through the water is the same.

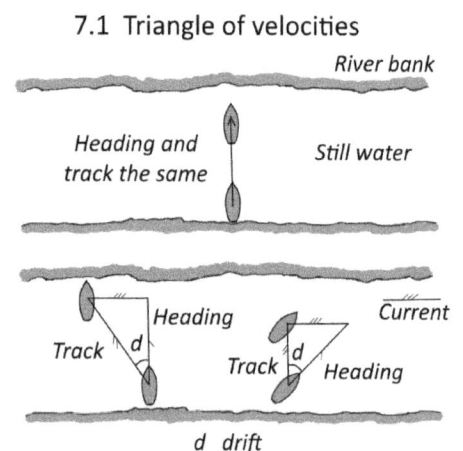

7.1 Triangle of velocities

In air-navigation, if we know the airspeed and heading (the direction we will travel through the air), and we know the wind-speed and direction, that makes two sides of the triangle and the angle between them. Then we can solve the triangle of velocities graphically or arithmetically and determine the third side of the triangle, the groundspeed and track (direction over the ground). In practice, we normally know the wind-speed and direction from a forecast and we know the intended airspeed and the desired ground-track. We need to calculate the heading and the groundspeed but of course, heading and groundspeed are not the same side of the triangle. The calculation can be done on paper, graphically or arithmetically but is awkward. The problem is usually solved using a Dalton computer which is a mechanical device which solves the triangle of velocities graphically and has a built-in circular slide-rule for the other calculations, or by using a specialised electronic device. The basic geometry is shown in figure 7.2.

Note that all the directions are relative to True North but the forecast or reported wind-direction is the direction *from* which the wind blows whereas the heading and track are the directions *to* which the aircraft is flying. Once we have the groundspeed we can calculate the time to fly the distance of the particular leg and work out fuel consumption and so on.

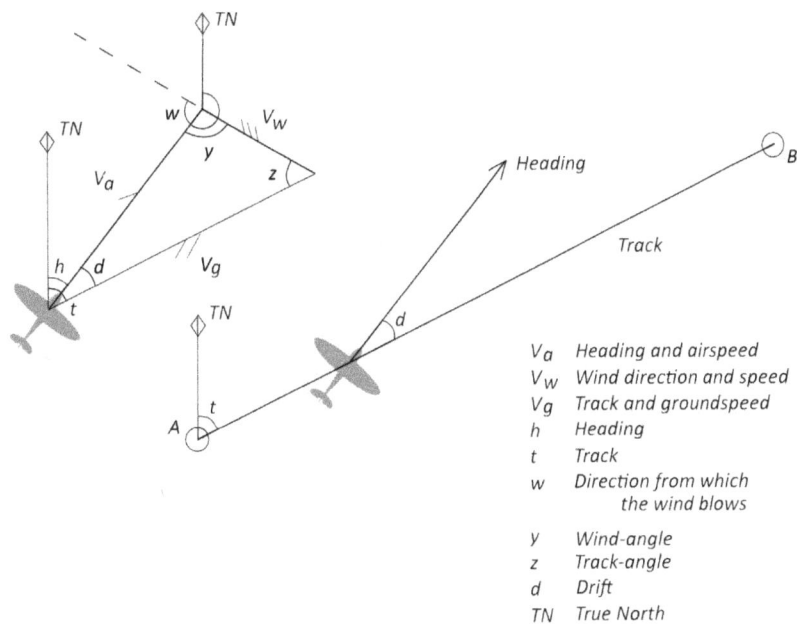

7.2 Dead-reckoning navigation

V_a	Heading and airspeed
V_w	Wind direction and speed
V_g	Track and groundspeed
h	Heading
t	Track
w	Direction from which the wind blows
y	Wind-angle
z	Track-angle
d	Drift
TN	True North

In dynamic soaring, the important thing to remember is that the triangle of velocities defines the relationship between air-velocity, wind-velocity and ground-velocity at any particular wind-angle and produces the drift angle, but the shape of the triangle is changing all of the time because the bird is turning. Also, it is not the triangle of velocities which makes the ground-velocity change; rather it is the force components acting on the bird or glider which cause the accelerations.

Map reading

Navigation is about the mental map that all creatures have in their minds; knowing where you are in that mental map, knowing where your destination is and knowing what direction to go to get there. You then compare your mental map with your actual surroundings in the real world and off you go. Taken to its absolute fundamentals, you do not need to know where you are (ask any sat-nav user!). What you need to know is: which way to go.

We think of a map as a sheet of paper (or an electronic device) with symbols on it depicting the real world in a diagrammatic way. A mental map is similar and begins in childhood with a map of a room and is then expanded by experience to include the neighbourhood and so on, ultimately being supplemented by those paper maps, for places we have yet to visit.

Map reading is a matter of comparing the real world you can see with the mental map or the paper map; then orienting yourself with the direction you need to go. Such orientation is achieved using some line feature to follow, a path or a river or a feature on your route which you can see from a distance. Or using some form of compass locked to a particular direction, for example the direction of magnetic north or some celestial feature combined with a time reference.

Animal navigation

Animal navigation is a real wonder of nature. How do creatures find their way around the world without a map or a compass or a time-piece; or without experience? Well, we can speculate that in their own way they have all of these things but in their minds rather than in physical form and that navigation is heavily controlled by instinct and presumably by genetic inheritance, particularly in relation to seasonal migration. It boils down to an instinctive reaction to a particular sensory stimulus. The stimulus could be something as simple as day-length. The reaction could be a desire to follow a magnetic field or a star pattern. Humans probably have similar abilities but we have learned to suppress our instincts and our sense of direction in favour of reliance on instrumentation. That said, ask any pilot who has a particular base of operations and a regular network of routes to fly, given good visibility, could he fly to any of his regular destinations without map, compass or clock. He will say yes. With a little experience and a view of the ground, you just know which way to go, even if the destination is thousands of kilometres away.

Experiments have been carried out to determine that particular creatures are sensitive to magnetic fields or the positions of stars in the night sky or the position of the sun in the day and so on. It is known that some birds fatten up before departing on migration and some feed en-route. They often wait for favourable winds before setting-off but whether they lay-off drift to achieve a particular track is uncertain. Certainly a hill-soaring bird does so to fly parallel with the hill in a cross-wind.

Albatrosses

Albatrosses go on foraging trips of several days or weeks travelling thousands of kilometres. Albatrosses are found in those parts of the world where the winds are both ancient and consistent; the North Pacific and the Southern or Antarctic Ocean. The large-scale wind patterns are typically circulations around seasonal high or low pressure areas or circulating around the Antarctic continent. By simply staying airborne using dynamic soaring and following the wind direction, the birds can follow the circulating wind patterns which will bring them back to their starting point.

This looks like a combination of systematic planning and random opportunism but the bird's instincts and their technique of dynamic soaring have evolved over millions of years during which time the wind has been blowing consistently, so that the wind, the dynamic soaring manoeuvre and the navigation are all linked together in its behaviour.

The analysis in chapter 9, of an albatross foraging expedition starting from the Hawaiian Islands, suggests that they can dynamic soar in any direction even directly upwind; so it would appear that they are not wholly dependent on the wind direction to get them to where they want to go. The thing is, we don't know what the bird's intentions are. Does it fly in a direction dictated by the wind or does it wait for a favourable wind according to its preference to fly in a particular direction? Is it going to a particular feeding ground or are there feeding grounds available at all points of the compass? If the bird ends up in the same feeding spot repeatedly, is that because it is deliberately flying there or is it the result of the same wind patterns taking it to the same location each time?

Faced with a featureless ocean, the bird does not need to know where it is, it only needs to know which way to go. A bird's mental map could be as simple (or as complicated) as a sense of where the sun is or should be in the sky. It is thought that birds have the ability to detect ultra-violet light, in their optical-sense; which might mean that they can see the sun and maybe the stars, even through clouds. The mental map may then comprise a sense of where the sun should be in the sky at a particular time at the roost. The bird's instinctive response is then to fly in a direction that moves the bird's view of the actual position of the sun towards the desired position in the mental map. Maybe.

Now we have meandered back to dynamic soaring, what is it that albatrosses actually do in dynamic soaring flight?

Chapter 8

What an albatross does

Observing dynamic soaring

Since mankind began exploring the great oceans we have been fascinated by the flight of the albatross. These huge birds are seen flying at low level over the sea with a characteristic, sinuous, undulating flight pattern mostly without flapping their wings. Research has shown they are capable of flying great distances at high average speeds with very little effort. Their flight manoeuvres are different to those used by land birds on soaring flights. So, what are they doing?

When you see an albatross in flight, you will see the bird, the ocean and the sky. You will not see the air, nor will you see the bird's airspeed, groundspeed, drift angle or its acceleration. Everything you can see, the bird, the waves, the swell, the water and you, are moving relative to everything else. The normal view is from the deck of a ship fairly close to the surface of the sea, so that the up and down motion of the bird is very obvious but less obvious is the extent of the left and right turns and their angle to the wind. Dynamic soaring is done at relatively high speed, for a bird. The birds will approach the ship quickly and disappear off towards the horizon, so that film clips of them doing classic dynamic soaring are rare and short. If the birds loiter around the ship in the expectation of a meal, there is no guarantee that they are still doing pure dynamic soaring because they will instinctively use any vertical movement of the air caused by the ship to sustain themselves in flight. If the observer's viewpoint is on land, there is even greater likelihood that the birds are hill-soaring.

People have made sketches of albatross tracks observed from the decks of ships but comparing these sketches with the actual tracks of albatross dynamic soaring made clearer by Global Positioning System (GPS) tracking, reveals a great disparity. Albatross dynamic soaring is often illustrated in diagrammatic form to support wind-gradient theory explanations but these diagrams rarely depict albatross flight realistically and are highly distorted to fit with the theory. In particular, they will depict the upwind and downwind turns as close to 180 degree reversals of direction with highly exaggerated vertical displacement and highly compressed horizontal movement, which is not what the albatrosses actually do.

Tracking seabirds

To explain albatross dynamic soaring, what is needed is a clear view of what albatrosses really do while dynamic soaring, using both video recording and data logging and the development of a theory which describes the forces and motions involved.

Albatrosses are routinely tracked by scientists using various devices, however the data is mostly of coarse resolution and unsuitable for analysing dynamic soaring. Naturalists are mostly content with data points several hours apart to determine where the birds are and what they are doing. Often the tracking does not even use GPS but rather, for example, it measures daylight and timing to determine latitude and longitude and wetness or temperature to determine whether the bird is on the surface or flying.

To properly analyse dynamic soaring, it will be necessary to measure, at fine resolution and simultaneously, both ground-velocity data (position, track, groundspeed and height) from GPS tracking, plus air-velocity data from airspeed and heading detectors together with wind-speed and direction data. So far, only the GPS data is available; simultaneous measurements of the air-data or the wind-data have not been made. Until such complete data sets are available any conclusions will depend on the assumptions made concerning the missing data. Nevertheless, using only the GPS data it is possible to estimate the wind-velocity and then calculate the heading and airspeed. We will then have a picture of dynamic soaring in terms of air-velocity, wind-velocity and height which will enable us to construct a realistic model of albatross flight.

Interpreting data logging of dynamic soaring

The data with which I am working is published in a paper entitled **In-Flight Measurement of Dynamic Soaring in Albatross by G Sachs et al AIAA 2010**. It appears in the paper as three diagrams comprising 30 seconds of data of GPS derived position, speed and altitude, covering three dynamic soaring cycles. Although there is a lot of bird-tracking data published on the internet, it is mostly position only; this is the only data-set I have found which includes height information. As published, the diagrams are highly compressed in their horizontal axes. To extract some useable data, the diagrams were expanded to give all three the same horizontal scale, as shown in figure 8.1.

8.1 Albatross GPS tracking data

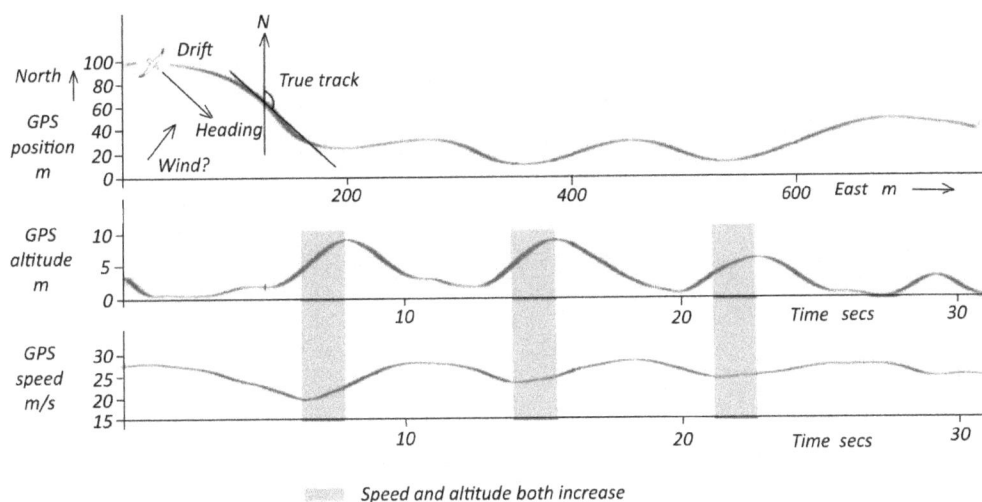

Speed and altitude both increase

The top line is GPS position, giving the true shape of the ground track, with North upward and East toward the right, giving the same scale in metres on both axes; the bird flies from left to right. True track in degrees (that is, track direction measured clockwise from True North or GPS Grid North) can be measured directly as the direction of the tangent to the line at suitable intervals, approximately 10m or half a second. The middle line is GPS altitude over the same horizontal distance but with an expanded vertical scale in metres. The lower line is GPS speed in metres per second. The middle and lower lines are the same horizontal scale as the top line but with units in seconds.

These data comprise relatively short windward turns compared to film of albatrosses in moderate winds where the windward turns are much longer. Also, the curvature of the windward turns corresponds to angles of bank which are steeper than those seen in the films with moderate winds. Therefore, I conclude that the wind experienced by the albatross being tracked by this data, is stronger than normal and the albatross is probably flying faster than normal to avoid excessively large drift angles and banking more steeply in the windward turn to avoid being blown too far downwind.

Without air-velocity or wind-data, the main point to be seen in this raw GPS data is that there are places, highlighted, where <u>both</u> altitude and ground-speed are increasing, indicating a net energy gain. This is very obvious during height gain. In fact, when kinetic and potential energy are summed (not illustrated here) the energy gain is seen to occur throughout the hump manoeuvre (the leeward wing-over turn). This is the main conclusion in the Sachs paper. What is not clear is how this energy gain relates to the wind-velocity and to the bird's airspeed and heading. This is not a simple trade-off of kinetic and potential energy, like a roller-coaster. In order to gain both ground-speed and height there must be applied forces and an input of energy.

Calculating the wind-velocity

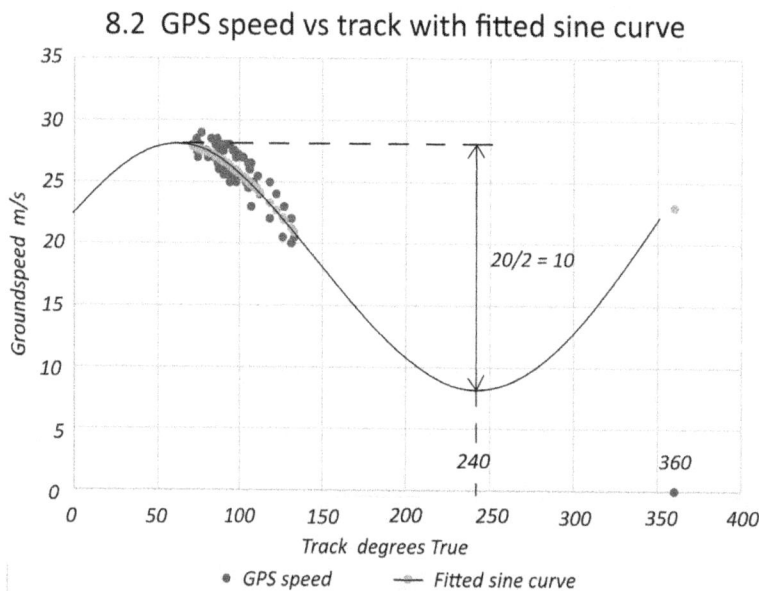

8.2 GPS speed vs track with fitted sine curve

There is a way of estimating wind-velocity from GPS tracking data which is explained in the next chapter. Suffice to say that groundspeed can be plotted against track and a sine curve fitted to it but it requires many more data points than we have here and a greater range of tracks. The result is shown in figure 8.2 giving an estimated wind of 240 degrees at 10 m/s. As you can see, with the relatively limited number of data points and the small range of tracks, there is great scope for inaccuracy. However, with the other information from the paper it is possible to make some informed guesses as to what the wind-velocity might be. Then, using that assumed wind and the raw ground-velocity data we can solve the triangle of velocities at each data-point at suitable intervals, to calculate airspeed and heading.

Deducing the wind-velocity

From the information given in the Sachs paper and general observation of albatross dynamic soaring flight we can make the following deductions:

1 - The hump manoeuvres on the altitude line are leeward turns, flown as wing-overs (belly to the breeze, if you like).

2 - On the position line, the left turns correspond to the hump manoeuvres and are therefore leeward turns (headwind to tailwind) and the right turns are windward turns (tailwind to headwind).

3 - Therefore, the wind is from the bird's right side, giving left drift and the bird's heading, in this case, is track plus drift.

4 - The data is from the vicinity of the Kerguelen Islands in the Southern Indian Ocean in the Roaring Forties where the prevailing wind is Westerly. Therefore, the wind is probably from between South-West through West to North-West.

5 - The bird's average heading is probably crosswind at 90 degrees to the wind.

6 - The bird's average groundspeed is about 27m/s. The wind-speed, at a first guess, could be half of this, say 13 m/s and the maximum drift would then be about 30 degrees. (Sine 30 = 0.5)

7 - From the position line, the average track-made-good is about 095 degrees, therefore, the average heading is 095 + 30 = 125 degrees True and the wind direction is approximately 125 + 90 = 215 degrees True (from the South-West).

8 - The assumed uniform wind is therefore 215 deg True at 13 m/s; compared with the calculated wind of 240 degrees 10 m/s.

Using either wind gives a similar result. (Using other values of wind velocity within +/-10 m/s and +/-20 deg gives similar curves but different values of airspeed. Radically different wind-velocities like due South, due East or due West give chaotic results).

Results with a uniform wind

Using 215 / 13 for the wind and the measured ground-speed and track data, we can solve the triangle of velocities and calculate the bird's airspeed and heading at each data-point at half second intervals which can then give an insight into the relationships between the airspeed, groundspeed, height and heading during the turns. The results are shown in the figure 8.3. The horizontal scale is time in seconds and the vertical scale is degrees, metres or metres per second as appropriate.

8.3 Albatross GPS tracking data
with calculated data using assumed wind

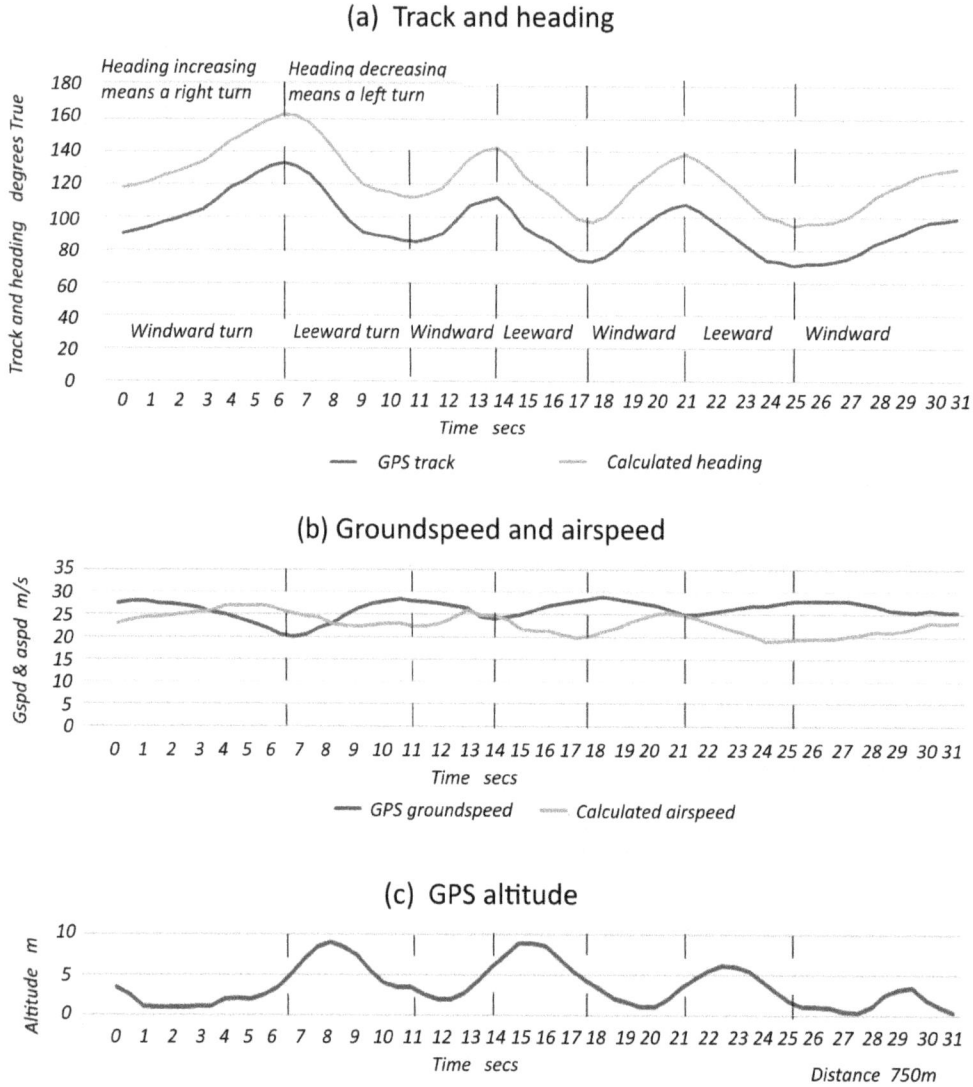

(a) Track and heading

(b) Groundspeed and airspeed

(c) GPS altitude

Heading and Track

The top line and the second-top line are respectively the bird's calculated heading and the GPS raw-data track in degrees measured clockwise from true North. This is not position information but purely numerical heading or track, versus time. The ascending lines represent increasing headings and tracks and are therefore right turns and are windward turns, turning from tailwind component to headwind component. The descending lines represent reducing headings and tracks and are therefore left turns and are leeward turns. The points where the lines change between ascending and descending lines, are the reversals of angle of bank and of turn direction. They are quite marked and show up much better than in the pure position line in figure 8.1.

The difference between the heading and track lines is the drift angle, which is approximately constant because we are assuming a uniform wind. The heading changes by about 40 to 50 degrees in each turn (crosswind plus and minus 20 to 25 deg) so that the drift angle does not vary by much.

The durations of the windward turns, at 4 to 6 seconds, are much shorter than seen in film of albatrosses in moderate winds, where the windward turns take 10 to 15 seconds. Also, while the average track is Easterly and the average heading is South-Easterly, the wind is estimated to be South-Westerly, compared to the prevailing Westerly wind at these latitudes. This suggests that the wind has backed (changed direction in an anti-clockwise direction in the Southern hemisphere) and possibly increased in speed due to a passing squall. Bear in mind that we are assuming a wind speed of 13 m/s which is about 26 kts or 30 mph.

Ground-speed and airspeed

The GPS speed line is GPS raw-data speed (groundspeed). There is a clear correlation between groundspeed and heading; the groundspeed and therefore momentum, reduces in the windward turn. In the leeward turn the groundspeed increases, despite the gain of height in the first half of the leeward turn.

The Airspeed line is calculated airspeed. The airspeed is not constant but actually it increases in the windward turn and decreases in the leeward turn. This, despite nearly constant height in the windward turn and the gain and loss of height in the leeward turn.

This is _not_ what is predicted by the wind-gradient theory in which the airspeed is supposed to increase in climbing and descending through the wind-gradient and decrease in level flight.

Altitude

The GPS Altitude line is GPS raw-data height above the GPS datum sea-level. Remember the horizontal axis is about 750m and the vertical scale is greatly expanded. During the windward turns the altitude is nearly constant. In the leeward turns there is a gain and loss of height, between about 1 and 9 m, due to the turn being flown as a wing-over. If the bird is skimming the surface in the windward turn, to take advantage of ground effect, altitude should vary somewhat due to passing waves and depending on the sea-state. That is difficult to see here.

Results with an assumed wind-gradient

The airspeed and heading data in figure 8.3 (above) were calculated using a uniform wind. However, if there is a wind-gradient, its effect will already be included in the GPS ground velocity data. To see a wind-gradient effect on the airspeed calculation, an assumption has to be made as to the nature of the wind gradient and we are back to theory again.

What is the structure of the wind-gradient? Well, certainly not a two-layer system as assumed by most wind-gradient theorists. Instead of assuming a uniform wind, the wind used can be modified to be uniform at and above an altitude of 10m and reduced at lower altitudes proportional to the logarithm of the bird's altitude, giving a quasi-exponential slope of wind versus height. The effect on the calculated data is seen in figure 8.4 (below).

Heading and track

The top line Heading is the calculated True heading; the next line GPS track is raw GPS True track. The difference between the heading and track lines is the drift-angle which is now variable due to the wind-gradient, reducing at low altitude due to the lesser wind.

Groundspeed, airspeed and height

The GPS speed line is the same raw GPS groundspeed as before. The Airspeed line is calculated airspeed including the effect of the wind-gradient. Average airspeed is almost exactly constant with most of the variation corresponding to the gain and loss of height in the leeward wing-over turns. The line is slightly flatter compared to the airspeed calculated with a uniform wind. This suggests that the effect of the wind-gradient is to reduce airspeed variability and thereby reduce drag losses in the leeward turn. (Aerodynamic drag follows a square law and the greater the variation of airspeed about a mean value the greater the total drag losses).

Again, there is no indication that airspeed increases when climbing and descending through the wind-gradient.

The GPS altitude line is the same raw height data as before with an expanded vertical scale.

8.4 Albatross GPS tracking data
with calculated data using assumed wind gradient

(a) GPS track and calculated heading using wind gradient

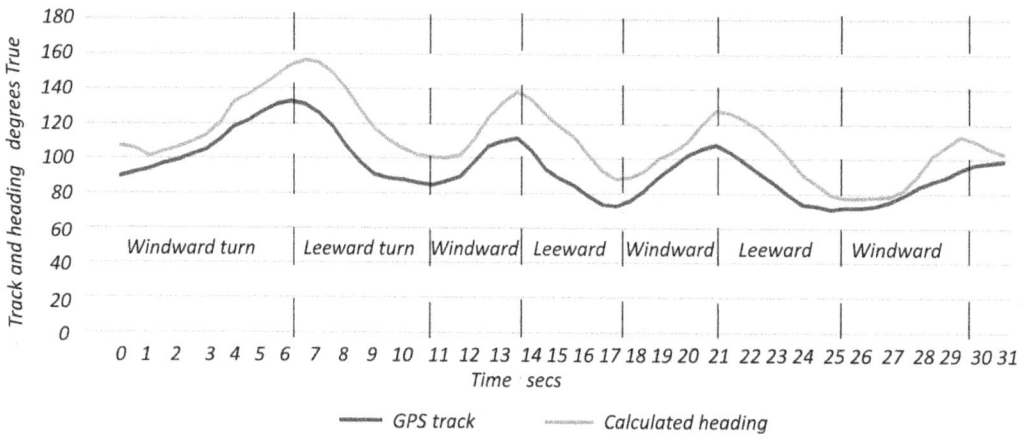

(b) GPS groundspeed and calculated airspeed with wind gradient

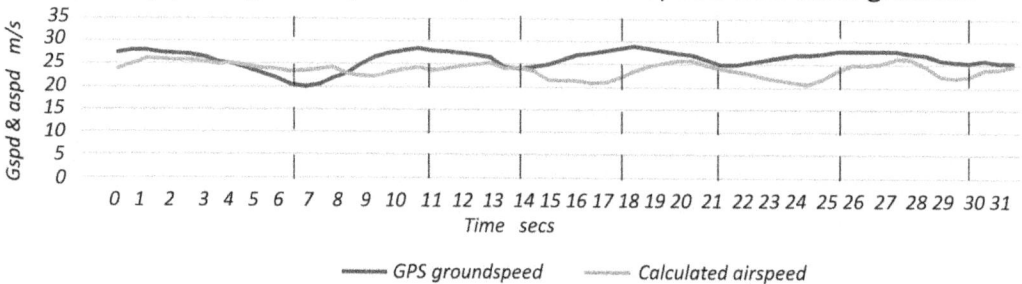

(c) GPS altitude

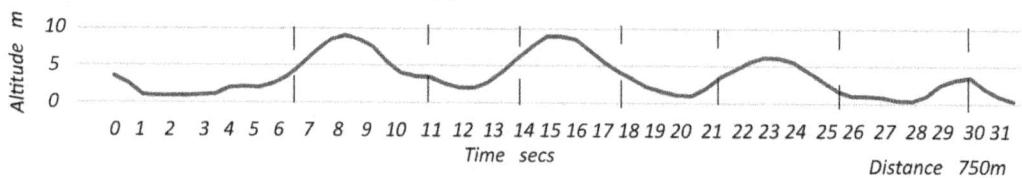

Summary

This data demonstrates what albatrosses are doing when dynamic soaring. They are not doing what the wind-gradient theory assumes; they are not flying 180 degree turns or circles. They are not gaining airspeed by climbing and descending through a wind gradient. They are not losing airspeed in level flight.

The GPS raw data are mostly smooth curves and do not show any sudden changes in track or groundspeed which might indicate a wind-shear layer. There is no sudden increase of airspeed as the bird climbs and descends through a supposed thin wind-shear boundary. (Although that does depend on the exact nature of the wind gradient or shear used in the calculation. If a sudden change of wind-speed is included in the assumed wind-gradient, you do get a marked change of airspeed in the result).

On the other hand, the data is entirely consistent with the Windward Turn Theory; which can explain what is seen in this data: why the windward turn is flat and is flown at approximately constant height close to the surface; how airspeed increases in the windward turn; why the leeward turn is a wing-over; how the bird gains groundspeed and momentum in the leeward turn and loses groundspeed and momentum in the windward turn and how the bird can maintain height and gain or maintain airspeed overall. The leeward turn is flown as a possibly low-G wing-over during which the groundspeed increases, even as the height increases and the airspeed reduces.

The bird's average airspeed, groundspeed and height is constant but there is a continuous expenditure of energy in the form of aerodynamic drag. The energy equivalent to aerodynamic drag losses must come from the wind. The Windward Turn Theory can explain how that exchange of energy and momentum between the wind and the bird takes place.

There is no absolute proof of anything here because the wind-velocity is an informed guess, although as mentioned before, radically different values for wind-velocity produce chaotic results which I regard as invalid. Nevertheless, it demonstrates that on the balance of probabilities, the wind gradient theory of dynamic soaring is unlikely to be the whole answer. Contrary to the Rayleigh Cycle, there is no significant gain of airspeed in the climb and descent and no significant loss of airspeed in the windward turn. In any event, the effect of the wind gradient is quite small because the bird never gets near to an up-wind or down-wind heading and the effect of the wind gradient is diminished by the angle off the wind. Also, the effect is negligible above a height of about three meters. Comparing the two diagrams shows exactly what the effect of the wind-gradient is – primarily a variation of drift angle.

The present global wind-patterns exploited by the albatrosses, were established during the movement of the continents to their present positions and the creation of the oceans in their current form over millions of years. Natural selection over the same time scale has favoured evolution of dynamic soaring for the albatrosses living in those parts of the world where winds are consistent and strong.

This is what an albatross actually does during dynamic soaring and this is what will be explained in chapter 10. Meanwhile, the next chapter will take us on a foraging trip with a Laysan albatross from the Hawaiian archipelago.

Chapter 9

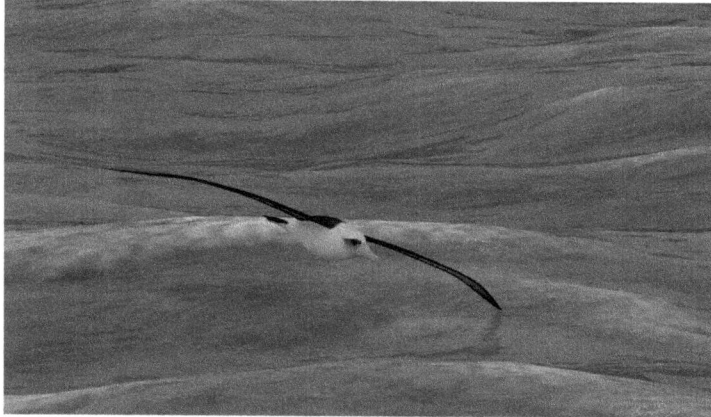

Albatross foraging

Analysis of Laysan Albatross GPS tracking data

In this chapter we will examine the flight path of an albatross to see whether anything can be learned about how albatrosses use dynamic soaring as part of their foraging behaviour. For scientists studying albatross behaviour, high resolution data or even GPS data is not necessarily needed. Data need only be collected every few minutes or hours rather than every second to determine where the bird goes and what it is doing; but that is not high enough resolution to analyse dynamic soaring. For that, resolution of one second or less is needed and, ideally height information as well. From time to time, high resolution data of albatross tracking becomes available and then we can get some insight into how dynamic soaring fits-in with the albatross' general foraging behaviour.

The Data

The dataset I am using here is one of several sets of GPS tracking of albatrosses and shearwaters, which was published on the website **datadryad.org** and comes from a research paper entitled: **Flight paths of seabirds soaring over the ocean surface enable measurement of fine-scale wind speed and direction by Yonehara et al, University of Tokyo 2017.** The paper was concerned with the calculation of wind-velocity by analysing seabird tracking data. This method could help to verify and fill-in the gaps in wind-velocity data derived from other sources such as ships, buoys and satellites. The data comprises approximately 46 hours of latitude and longitude coordinates recorded at one second intervals but with no height information.

Calculating wind-velocity

The method of calculating the wind-velocity works like this: Using the difference between successive GPS latitude and longitude positions at one second intervals, the groundspeed and the track direction relative to True North are calculated. On a scatter-graph the data are plotted with groundspeed on the y-axis against each possible track from 0 to 360 degrees on the x-axis.

When airspeed and wind-speed are both uniform, the variation of groundspeed versus track is an approximate sine-curve. The lowest and highest values of groundspeed occur at the exactly into-wind and down-wind tracks respectively therefore this gives the direction of the wind. During a full 360 degree circle, the groundspeed varies by double the wind-speed therefore the wind-speed is half the difference between the greatest and least groundspeeds.

Since, in the natural world, neither a bird's airspeed nor the wind-speed are exactly uniform and typically they are not flying circles, the distribution of the groundspeed values is a 'smeared', approximate and partial sine curve, which will only be a complete curve if data points are obtained for all tracks from 0 to 360. Furthermore, the data distribution is slightly 'pinched' at the point where the groundspeeds are greatest because the rate of turn and the rate of change of speed, is also not uniform. In practice, a 'partial, smeared' curve is obtained by plotting data from a given time period, say 300 data points from a 5-minute section, to which a clean sine curve can be 'fitted', and that curve is used to get the average wind-velocity. (Despite appearances, the graph is not a pictorial view of the dynamic soaring manoeuvre. For instance, the section from the least speed to the greatest speed is the leeward turn).

Figure 9.1 is an example of such a graph comprising 300 data-points at one-second intervals, plus the fitted sine curve. The numbers in the title refer to the number of seconds into the data set. The gap in the data means that there were few tracks between about 270 and 310. The lowest point of the curve is at a track of about 130 degrees and 9 m/s. In other words, the bird was tracking South-East when heading directly into the wind. The highest point is at 310 degrees and 17 m/s. The difference between the highest and lowest groundspeeds is 8 m/s. Therefore, the wind-velocity is 130 degrees True at 4 m/s. This exercise is repeated for any particular location in the total data set. Notice that this method uses ground-velocity data; the airspeed of the bird is not needed.

9.1 Wind velocity calculation
Groundspeed vs track

Data-set 144000-144300

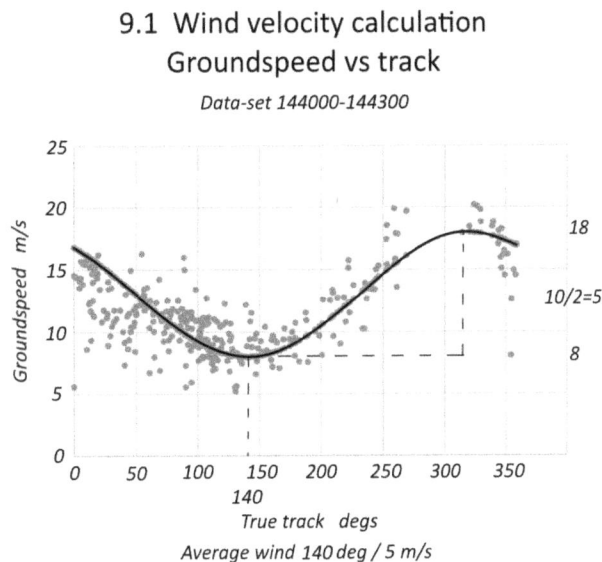

Average wind 140 deg / 5 m/s

The wind-direction, 130 degrees from the South-East, is the wind FROM which the wind is blowing, which is the meteorological and navigational convention.

The problem with using albatross dynamic soaring for the specific purpose of calculating the wind, is the fact that average dynamic soaring tracks on a scale of kilometres are relatively straight and a single dynamic soaring manoeuvre, comprising a single windward turn and a single leeward turn, on a scale of hundreds of metres, has a relatively narrow range of tracks: typically, only sixty degrees. Nevertheless, five minutes of data with sufficient variation of track can give a large enough data set and, given the relatively short time period, presumably, relatively uniform winds; it is then quite easy to manually fit the sine curve to the data.

Raw data

Before looking at particular points of interest, what can we do with the raw position and speed data? First, the overall track can be plotted on a rectangular grid to give a true scale representation of the track, as shown in figure 9.2. Second, the wind-speed can be plotted against time as shown in figure 9.3. Third, the ground-speed can be plotted against time as in figure 9.4.

The Ground Plot

We can see here 46 hours of data comprising 42 hours in the life of a Laysan albatross, travelling from bottom to top, plus four hours of data from before the tracking device was attached to the bird. (Figure 9.2). The wind-velocity is indicated by the arrows and in degrees True and meters per second, at one hundred kilometre intervals. The circles are the locations of points of interest that we will look at later.

We don't know whether the bird is male or female or whether an experienced adult or a juvenile, so let's just call him Phil. Hi Phil. The journey begins in February 2014 at Ka'ena Point on the North West corner of Oahu in the Hawaiian Islands.

Once the GPS tracker is attached, Phil sets-off in a determinedly Northerly direction. In 42 hours Phil will travel 1000 kilometres to the North. This is surprising because the wind is a light Northerly for the first 36 hours. Phil's main soaring technique - dynamic soaring - is theoretically most energy-efficient when flying on crosswind headings but clearly upwind dynamic soaring in light winds is also a practical proposition.

To achieve an overall journey of just over 1000km, the total ground distance flown is 1747.403km. This is the sum of all of the one second increments and may not be exact. It is the combined effect of the sinuous nature of dynamic soaring at the small scale of a few hundred metres and the meandering flight path at the larger scale of hundreds of kilometres. This apparent inefficiency is offset by the energy-saving of dynamic soaring at relatively high speeds using only the energy of the wind.

9.2 Ground plot Data-set 5 - 166620 = 46.28 hrs

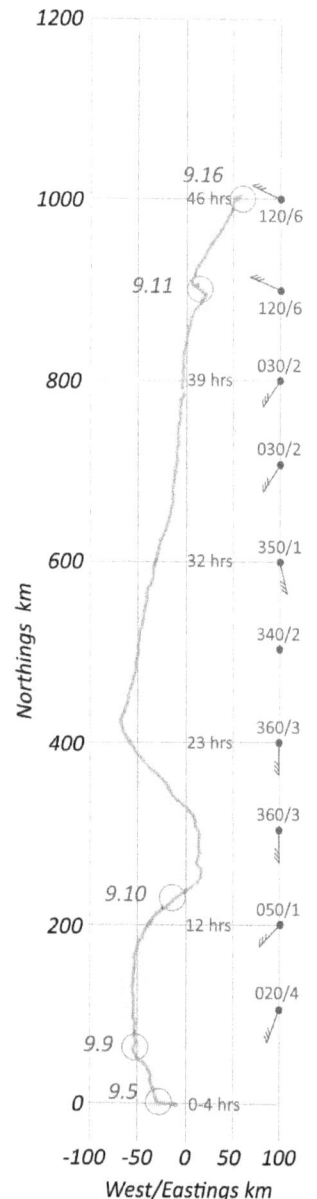

Why is Phil heading North? If not led by the wind or smells carried on the wind, then it may be that Phil is heading for a preferred feeding ground or even following another bird. Maybe one day scientists will fit a camera along with the tracking device.

Wind-velocity

In the next diagram, the wind has been calculated at one hour intervals and plotted as a bar graph against time in hours and in seconds. The first five hours of non-flying data are excluded from this. (Figure 9.3).

We might expect that the wind strength at low levels would reduce at night with cooler temperatures and less atmospheric turbulence and mixing. To see whether there are any diurnal effects, the two markers are spaced 24 hours apart but arbitrarily placed. The wind data appears to show two reduced wind-speed periods but it is difficult to exactly fit the two 24 hour spaced markers to demonstrate any diurnal effect. Bearing in mind that the two markers are also 500km apart, the variation of wind is probably a combination of geographical, meteorological and diurnal effects rather than purely an indication of night fall.

9.3   Wind-speed at 1 hour intervals

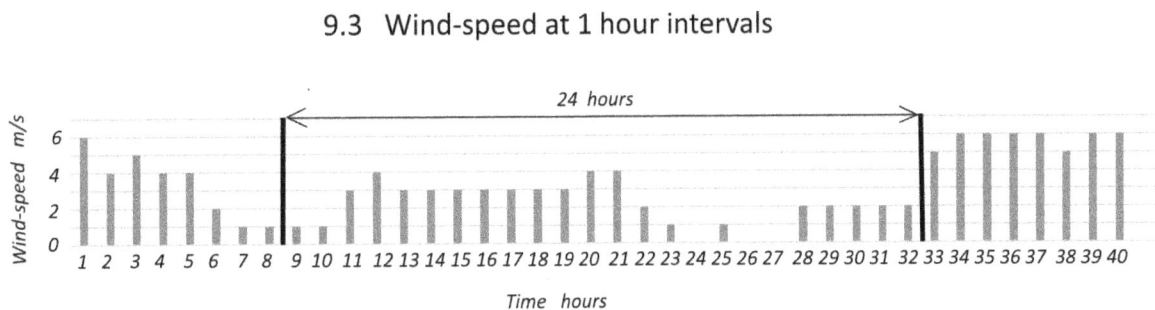

Ground-speed

On the Ground-speed graph below (Figure 9.4), the first four hours are mostly to do with scientists switching-on the tracking device and carrying it by car and on foot to the roosting site prior to attaching it to Phil. The rest of the data after about 18,000 secs is pure albatross flighting or alighting.

While airborne, the Phil's groundspeed varies to different degrees. The variation of groundspeed is either due to a variation of airspeed or a variation of wind-speed or a wider or narrower range of wind-angles used, in other words, more or less turning. The variation of Phil's groundspeed is most likely to be due to the strength of the wind. The greater amplitudes of groundspeed do seem to correspond to the increasing wind-speeds during the latter part of the journey, while the smaller amplitudes of ground-speed appear to correlate to the lesser winds during the presumed night-time periods. If the reduction of wind-speed is an indication of night fall, there is no sign that Phil is sitting it out. When the wind reduces to near calm at about 23 hours flight time, Phil alights more frequently but never for long periods. Is he feeding or resting or sleeping?

9.4 Groundspeed

Data-set 11000-166620 = 43.23 hrs
(including 3.23 hrs pre-flight)

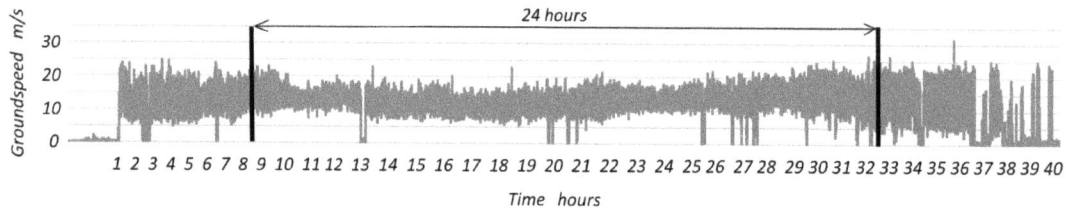

Flying time

The groundspeed graph indicates when Phil is flying and when he has alighted. Albatross flying speeds are approximately 10 to 20 m/s therefore, a ground-speed of less than about 4 m/s shows when he has alighted.

Phil was airborne for no less than 36.89 hours out of 41.96 hours of his journey; that's 88% of the time. During the first period of 37.2 hours Phil was airborne for 36 hours; that's 96.75% of the time. During that first period he alighted 20 times for an average of approximately 2.5 mins each time.

For the shorter second period of 4.76 hours Phil was airborne for 0.91 hours, which is 19.1% of the time. In other words, during the second period he was on the surface for 81% of the time.

Foraging strategy

After 32 hours the wind gets stronger from the South-East and the variation of ground-speed becomes greater. Then at 37 hours, North of 700km, there is a change in the groundspeed pattern. Phil starts making periodic 90 degree turns upwind or downwind. This could be a sign that Phil's track is intersecting a scent trail and he is turning to follow it. A scent trail will presumably drift downwind. Finally, at the Northernmost point of the track it looks like Phil has found what he is looking for and he spends most of the time on the surface.

Maybe we can conclude that Phil has reached his feeding ground but how did he achieve that? Did he go to a known location or did he follow a scent trail upwind or something else? A valid foraging strategy would be to fly upwind when a scent trail is detected and fly crosswind if the scent trail is lost. Turn upwind again when a scent trail is detected again. We can look for evidence of this by zooming-in on selected sections of the data.

Taking a closer look

Now let's look at things in more detail. The first four hours of the data are nothing to do with Phil the albatross. The GPS tracker is switched on and then transported to the roosting site. After the tracking device is attached, Phil wanders around for an hour or so within a few tens of metres.

Taxying, pre-flight checks and take-off

The next diagram (Figure 9.5) takes up the story and covers about 5 minutes of time but is only 30 metres from one side to the other. The location is the roosting site at Ka'ena Point on Oahu. Phil is depicted to scale. After shuffling around for a bit at point A, he lurches forward to B, going 15 m over the ground in 3 seconds, maybe 30 m through the air with a headwind.

The ground is flat here and the wind relatively light; Phil is surely flapping! He lands at B and shuffles to C, 10 m to his left. Is he practising flapping; getting ever lighter on his feet? Is Phil a juvenile practising for his first solo? An experienced bird would surely not be messing about like this.

Finally, not with a lurch but a launch, facing the Northerly breeze, striding out and flapping gamely, in seven seconds from C to D, Phil accelerates to 10 m/s; airborne and free!

Take-off

Next, we see the take-off again as Phil achieves full flying speed from point C to point E. (Figure 9.6). The wind is from the North-north East, 030 degrees 5 m/s, about 11 miles per hour. The scale bar is 50 m. The groundspeed trace shows 87 seconds of shuffling about on the ground and 11 seconds of flight. After achieving flying speed, Phil immediately bears off to the left to achieve a cross-wind heading. An experienced bird would immediately go into dynamic soaring mode but Phil does not. If he is a juvenile, he has never done dynamic soaring before.

9.5 Pre-flight and take-off
Data-set 18685-18985 = 5 mins

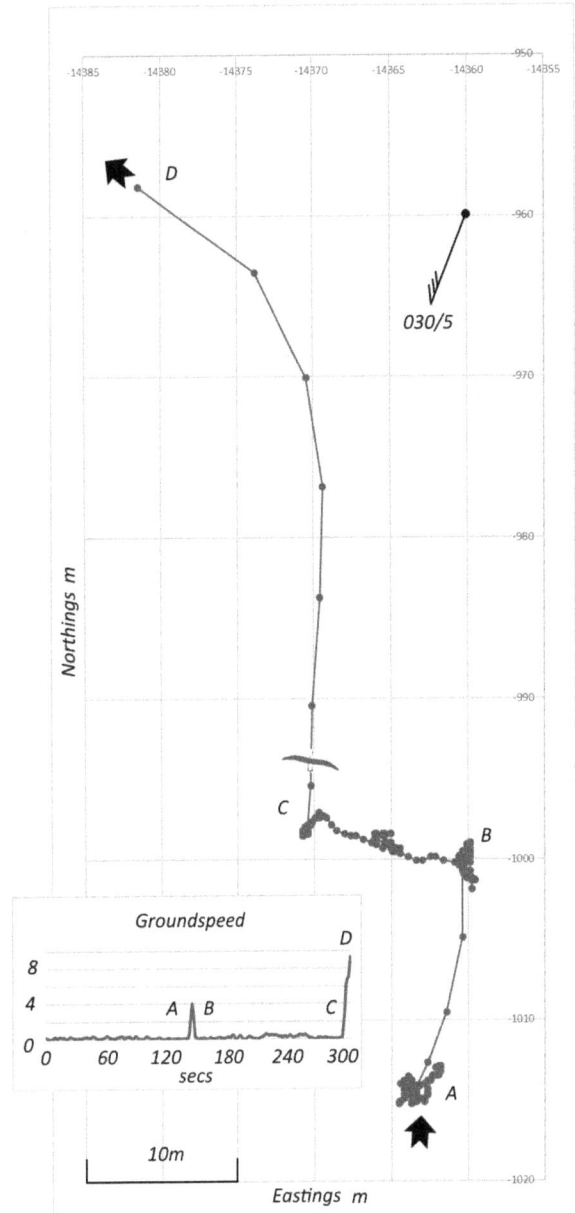

9.6 Take-off

Data-set 18890-18988 = 98 secs

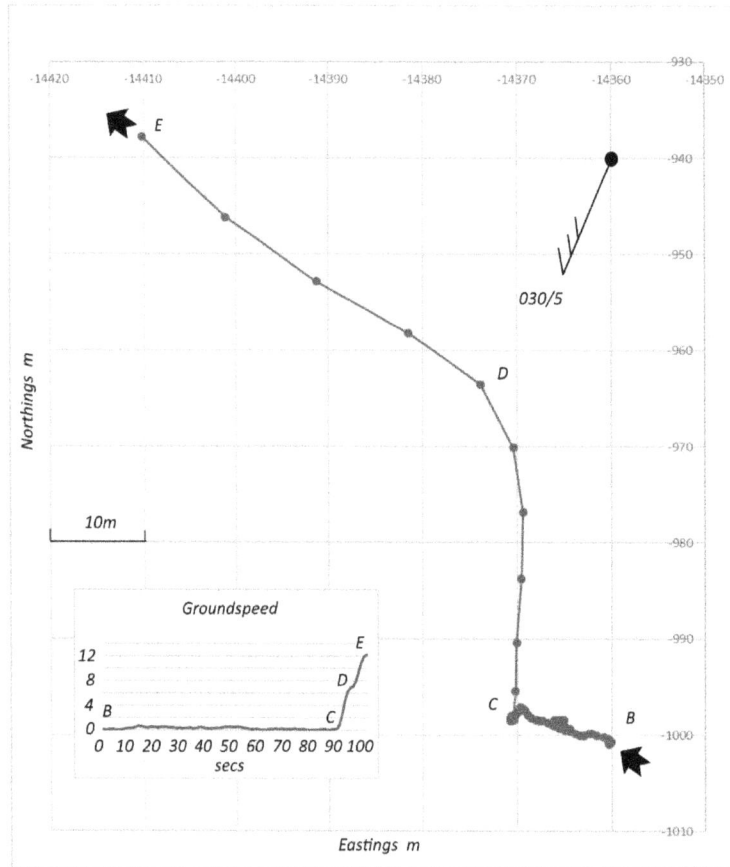

The first 5 minutes of flight school

The next diagram is a larger scale - note the scale bar of 500 m - but is only 5 minutes of data. (Figure 9.7). The wind analysis chart shows a wind of 030 degrees 5 m/s. Phil's track is relatively straight from E to F, although still snaking +/- 10 to 15 degrees. Groundspeed is 13 m/s +/-2 m/s with a wavelength of about 8 seconds. This is too slow to indicate flapping frequency but it probably is flapping; the variation of groundspeed corresponds to the variation of track which suggests constant airspeed but obviously we cannot see if there is any variation of height.

Suddenly, at point F after only a minute of flight, something clicks in Phil's brain. He locks his out-stretched wings and changes to a more sinuous motion. The wavelength of the motion increases to about 12 seconds and groundspeed increases to 15 m/s +/-5 m/s. The groundspeed graph shows the change of flight-style at point F. This is easier than flapping and he can sense through his tube nostrils that he is sustaining his airspeed. This is dynamic soaring!

The flight path from F to G gives a greater variation of ground-speed due to the greater variation of track direction. Where the data points are more widely spaced indicates greater groundspeed and therefore more downwind headings. From F to G, Phil practises curves and circles. The pattern is not well-ordered although it is crosswind which is the most efficient dynamic soaring orientation. He will keep this up for an hour before alighting briefly, maybe to feed, maybe to rest. Certainly, if Phil is a beginner he is wise to avoid splashing down near the roost in the fledging season. You never know what might be lurking below the surface! Sharks are known to patrol the sea near albatross roosting sites at fledging time.

9.7 Flight school

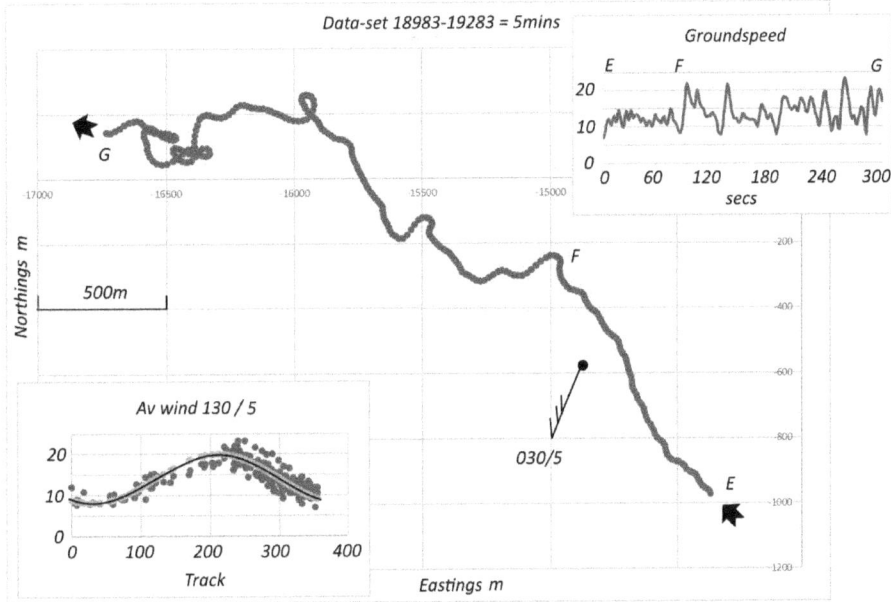

Data-set 18983-19283 = 5mins

The first touch-down at sea

Moving on, (Figure 9.8) Phil is 13.5 kilometres North of Oahu, after an hour of flying upwind into a North-easterly breeze; here we can see seven and a half minutes of Phil's track on a grid of 100m squares. Phil alights at point A for about 5 minutes before setting off again. Although most of the distance of this particular plot is in flight, most of the time, from A to C, is spent on the surface.

9.8 Touch down

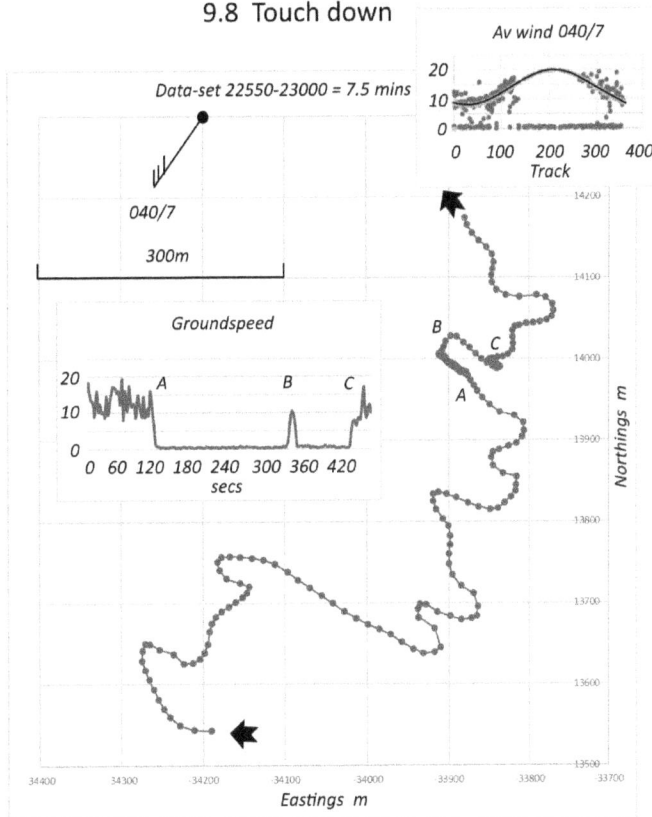

Data-set 22550-23000 = 7.5 mins

Getting the hang of it

The next glimpse of Phil's progress, after nearly five hours, sees him 70 kilometres North of Oahu, doggedly pushing-on into a light North-easterly. (Figure 9.9) His flight-path is undulating but still somewhat irregular, his average groundspeed is approximately 13m/s.

9.9 Getting the hang of it

Data-set 33000-33300 = 5 mins

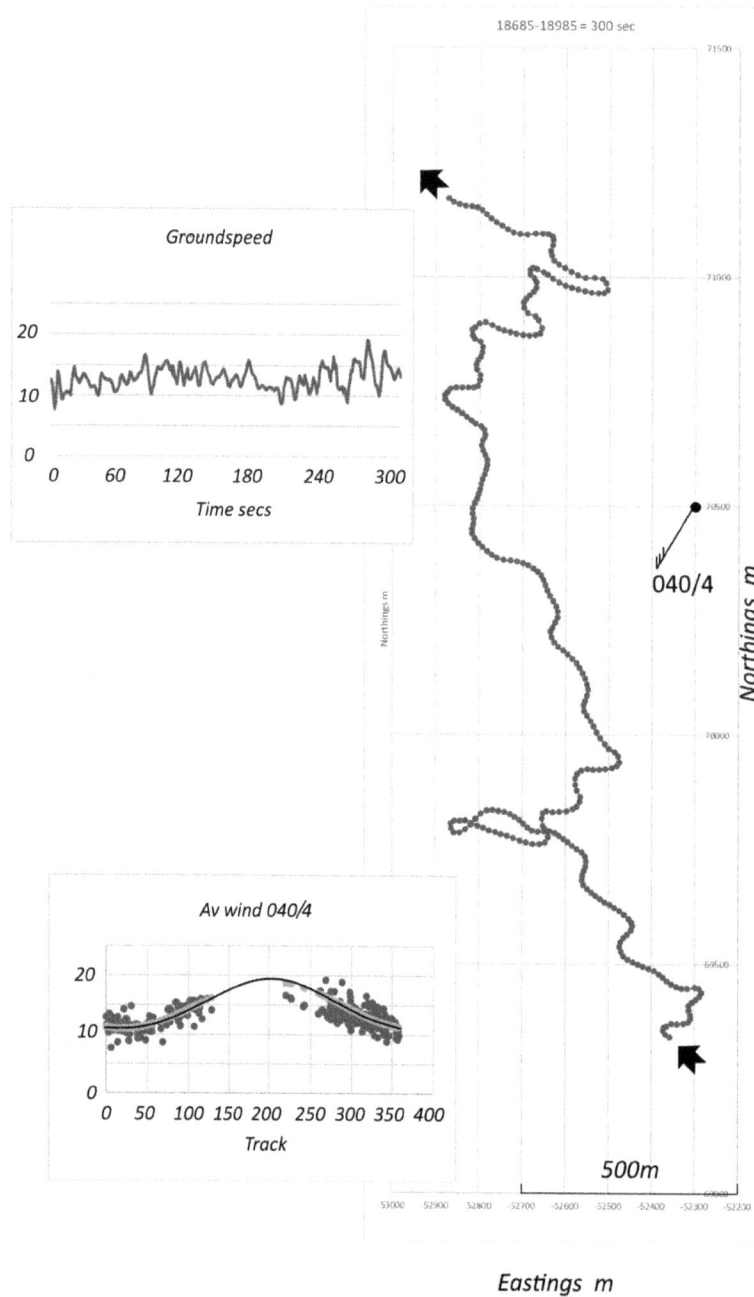

Night flying

After just under 11 hours of flight time, Phil is 230 kilometres North of home and is tracking North-East at around 14m/s. (Figure 9.10).

9.10 Night flying

Data-set 53000-53300 = 5 mins

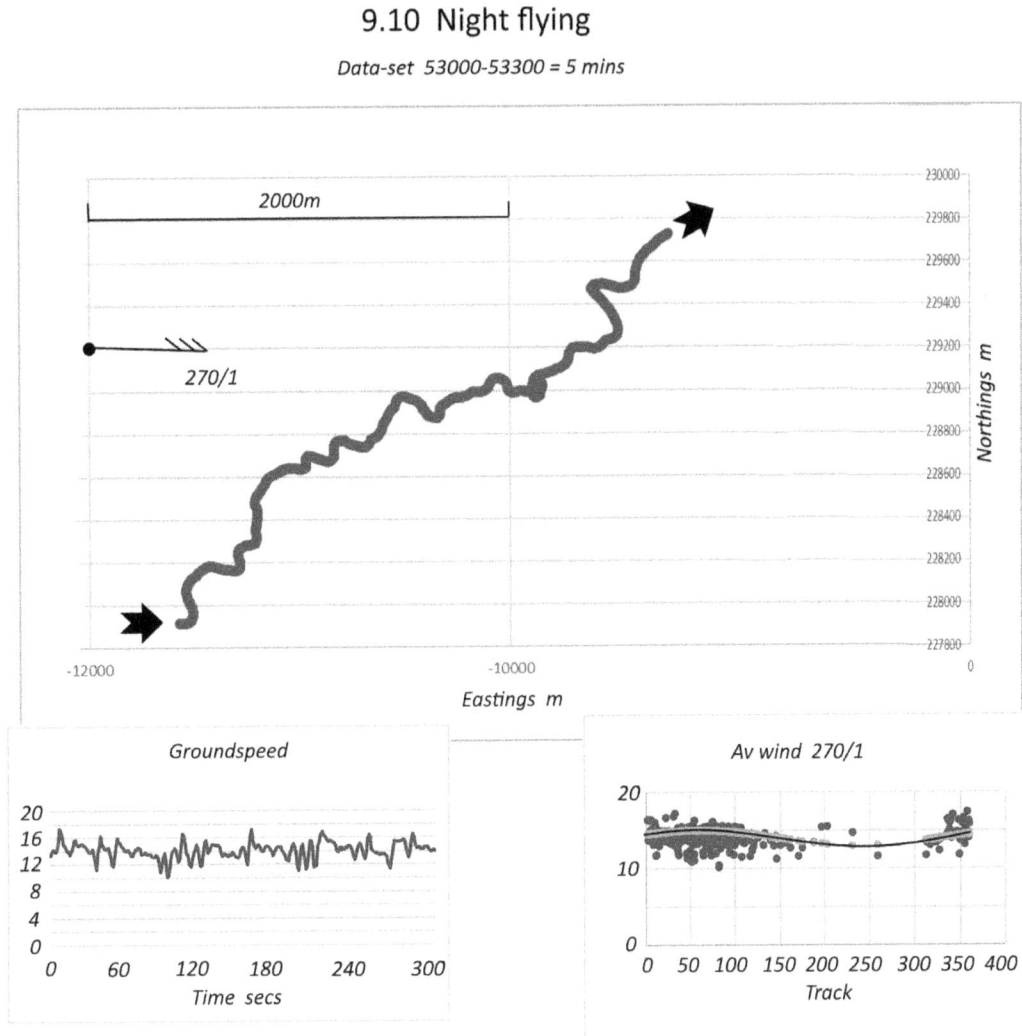

The wind has died away to a negligible westerly or light and variable. If we assume Phil was tagged around midday on the first day, then it is probably night but possibly moon-lit or star-lit.

Phil's ground speed has increased slightly because he now has a slight tailwind. The groundspeed record looks different compared to the previous diagram. There are periods of relatively little speed change indicating relatively straight flight and probably flapping. Then there are intervals of greater speed change indicating more turning and probably dynamic soaring. Considering the very light wind, it could be that there is more flapping going on here. This 5 minute section is 3043m as the crow flies but is 4200m as the albatross flies. The energy saving of dynamic soaring makes it worthwhile travelling 38% further and taking longer to achieve the same distance.

Turning upwind and downwind

At 880 kilometres North of Oahu, the wind is now from the South-East and Phil is tracking North East with a steady 5m/s crosswind.

During the 3.3 hours of this extract, covering about 55 kilometres, clearly something different is happening and it is worth taking a closer look (figure 9.11).

The first two points of interest A and B do not look very different in the Ground plot view but in the Groundspeed chart, section A is a flight sequence and section B is a surface event. Section A is shown in the next diagram.

9.11 Upwind and downwind

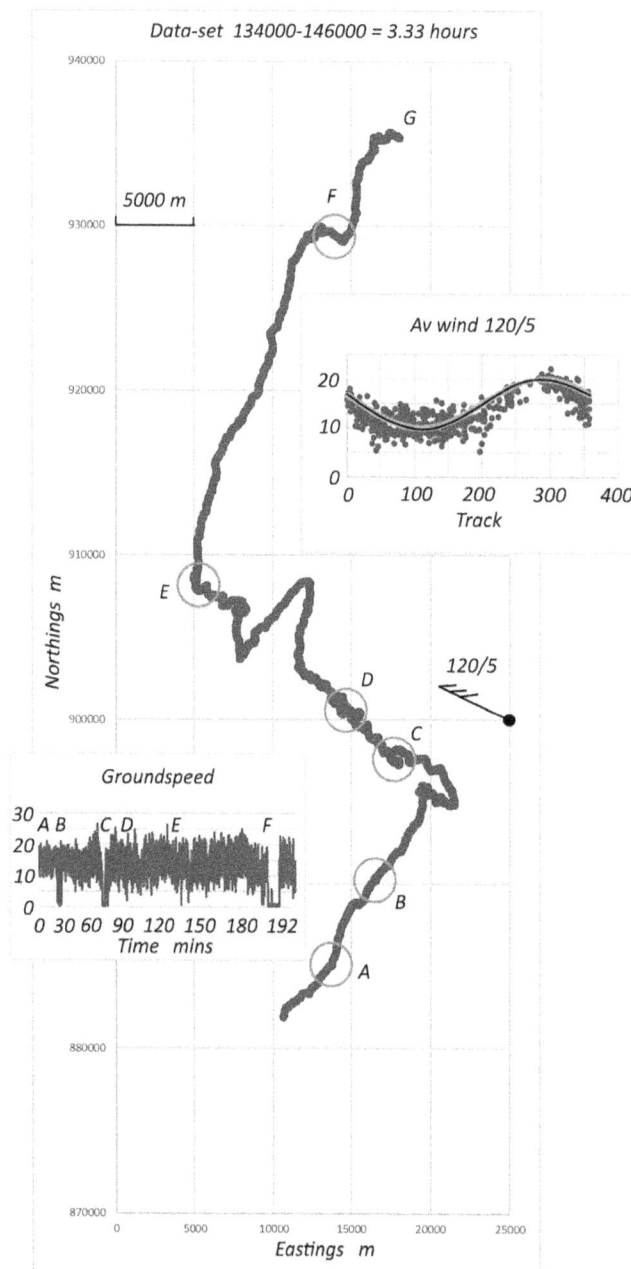

Data-set 134000-146000 = 3.33 hours

Av wind 120/5

Groundspeed

Dynamic soaring

The first event A, shown here as A1 to A2 (figure 9.12), illustrates 2 minutes of the relaxed and easy grace of classic dynamic soaring; an average cross-wind heading with long, windward turns of 5 to 10 seconds to the right and quick leeward turns to the left at an average groundspeed of 15 to 20m/s.

9.12 Dynamic soaring

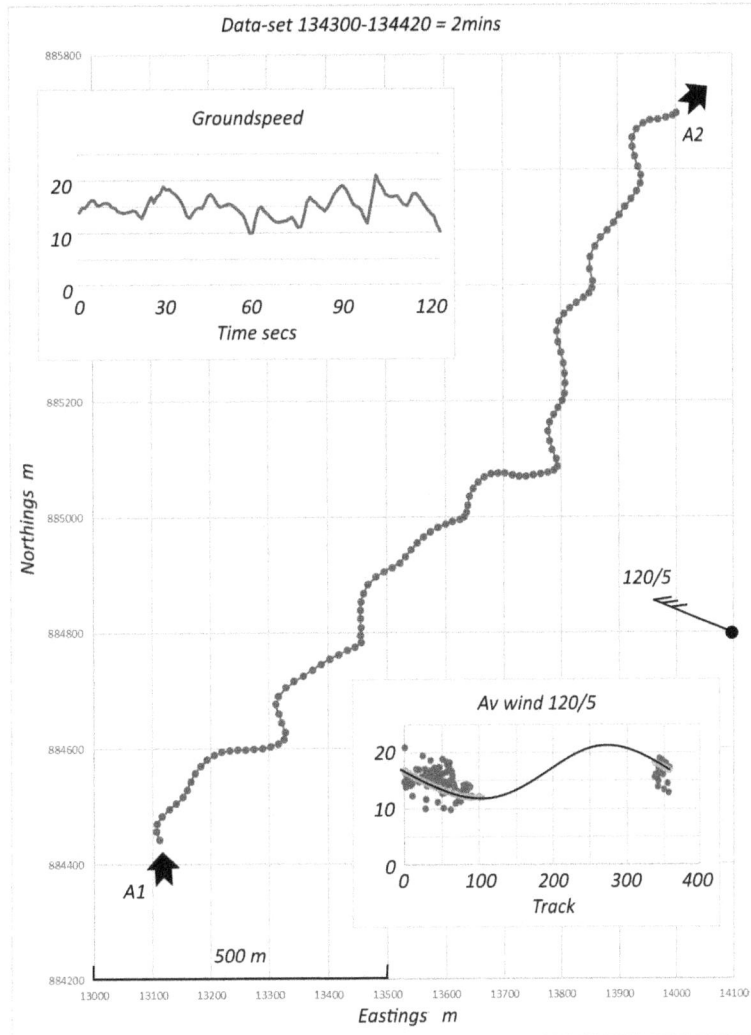

This looks much more orderly, smoother and more efficient compared with Phil's efforts a day and half before. Is this an illustration of how a juvenile albatross masters the art of dynamic soaring during its first two days on the wing?

Tacking upwind - gybing downwind

Now we are at point C in figure 9.11, shown as points C1 to C5 in figure 9.13. The wind is 5 m/s from the South-East. Suddenly at point C1, Phil takes a turn to the right and heads upwind for a couple of kilometres towards the South East. With apparently nothing found, or at least no reason to alight, at point C2 he then doubles back downwind to the North-West and picks up the trail again at point C3. But note he does not fly directly downwind. Rather, he zig-zags or gybes downwind like a sailboat, suggesting that while dynamic soaring works crosswind as A1 to A2 (with a downwind drift element) or upwind as from C1 to C2, it does not work so well for tracking directly downwind. This is explained later in chapter 15.

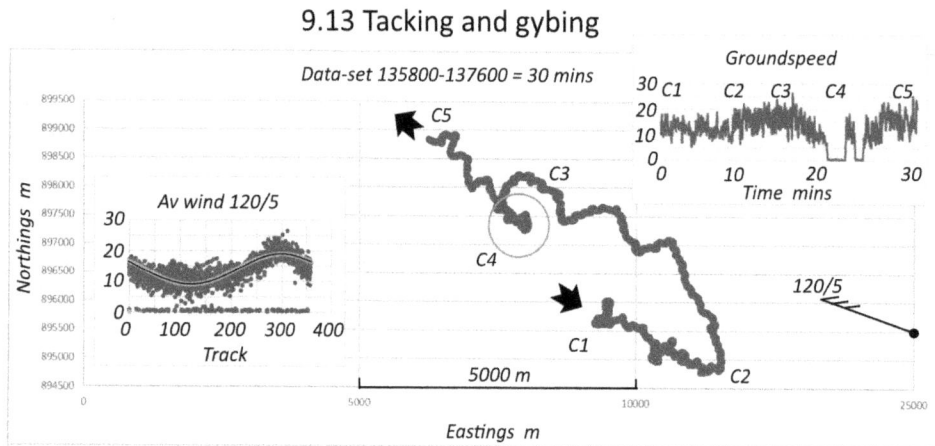

9.13 Tacking and gybing

Zig-zagging downwind

Next is an expanded version of the section from C2 to C3 (Figure 9.14). In the Windward Turn Theory of dynamic soaring, it is indeed difficult, if not impossible, to find a solution that allows a direct downwind flight path. Upwind or crosswind flight path solutions are relatively easy to find but downwind tracks are best achieved with a variation of the crosswind technique which can be flown at approximately 1G load-factor but at about 135 degrees to the wind direction. In this case, the zig-zag section from C2 to C3 comprises five legs, each with about five dynamic soaring manoeuvres on each leg and yields an average groundspeed of about 15m/s. A direct downwind distance of 3000 m in 5 minutes gives an average speed of 10 m/s which is faster than the 5 m/s wind and more efficient than simply meandering around at random and drifting downwind at the speed of the wind, as seen later. Looking at the downwind leg again, from C to E in figure 9.11, we can see the zig-zag pattern repeated in a fractal manner at scales of 100m 1000m and 5000m.

9.14 Zig-zagging

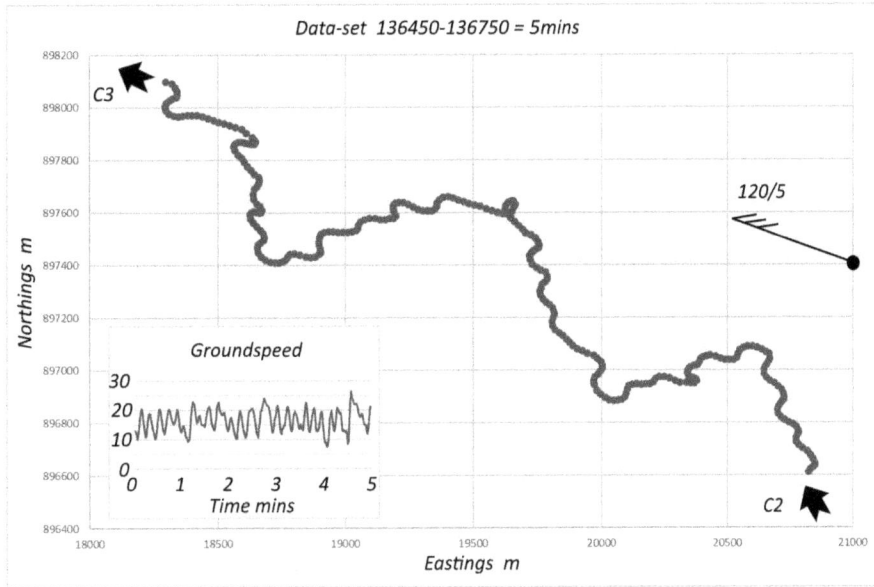

Data-set 136450-136750 = 5mins

Meandering downwind

9.15 Meandering downwind

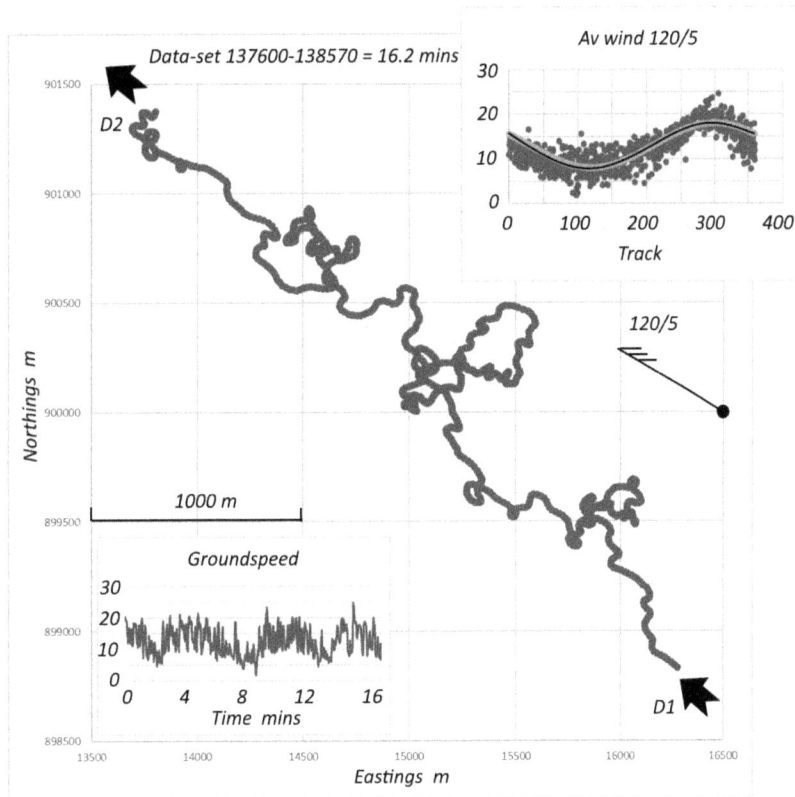

Data-set 137600-138570 = 16.2 mins

Next is the section at D in figure 9.11, shown as D1 to D2, in figure 9.15, illustrating a different downwind flight pattern. Instead of the well-ordered zig-zag track from C2 to C3, the flight path here looks more chaotic with a groundspeed varying from 10 to 20 m/s; covering a direct-line distance from D1 to D2 of 3200m in 16.2 minutes, at an average point to point speed of only 3.3 m/s which is less than the 4m/s speed of the wind.

Phil appears to be interested in this particular patch of sea, moving slowly in the downwind direction, but does not appear to alight. Clearly, this is a less efficient way of travelling dowwind compared to the zig-zag method and after about 16 minutes Phil reverts to the more efficient gybing pattern from D to E (Figure 9.11).

During 1.8 hours, it appears that Phil has investigated a definite line from C to E, about 20 kilometres long, aligned approximately with the wind. This could mark a scent trail drifting downwind (figure 9.11).

Finally, at point E he breaks off and heads North again.

Journey's end

About 40 hours into his journey and 1000 kilometres North of Oahu, Phil seems to have mastered his craft. (Figure 9.16). The wind remains at about 5 m/s from the South-East; Phil is tracking North-North-East. From J to K we see classic dynamic soaring flight. The flight path is undulating on a scale of a hundred metres but is fairly straight on a scale of 1000m. Maintaining a sinuous cross-wind heading, the resulting track is slightly downwind, according to the drift angle. An average triangle of velocities is added to illustrate the three velocity vectors and the angle of drift.

9.16 Journey's end

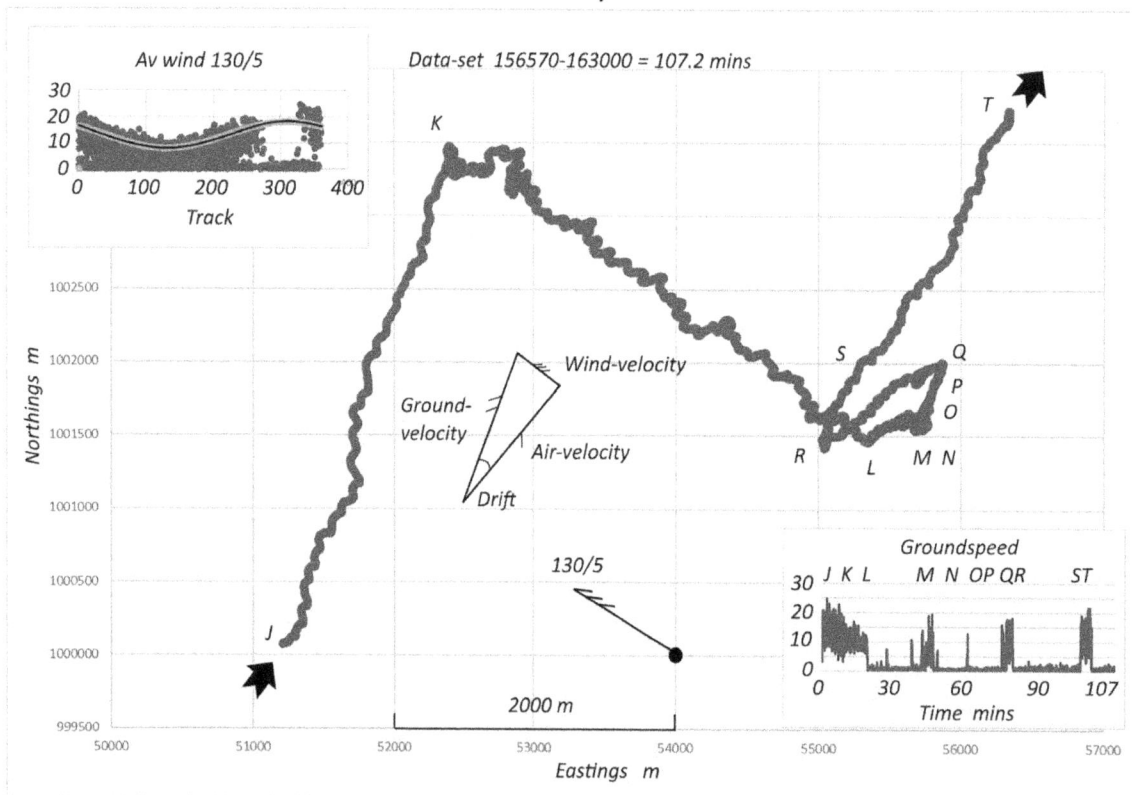

At K, Phil takes a sharp right turn through 90degrees and begins to head upwind to the South-East. Has his cross-wind track once again, intercepted a scent-trail drifting downwind?

Comparing the crosswind leg J to K with the upwind leg K to L, the character of the undulations changes subtly, becoming more compressed with the headwind. On the ground-speed graph, the ground-speed from K to L is visibly less than from J to K. The dynamic soaring technique is slightly different when going upwind compared to crosswind. According to the Windward Turn Theory, the angle of bank and load factor are greater during upwind dynamic soaring, requiring greater effort from the bird.

At point L, Phil alights and now spends most of his time on the surface with brief flights. The circled area from L to S is expanded below. The wind speed has picked up a bit to 8 m/s but still from the South-East. (Figure 9.17).

Between L and S, Phil spends a lot of time on the surface with brief linking flights. From Q to R is such a linking flight. We can see the classic crosswind dynamic soaring flight; long windward turns at a slow rate of turn linked with short, quick leeward turns. Phil alights at R and pursues another line of interest to S, normal to the wind.

9.17 Finale

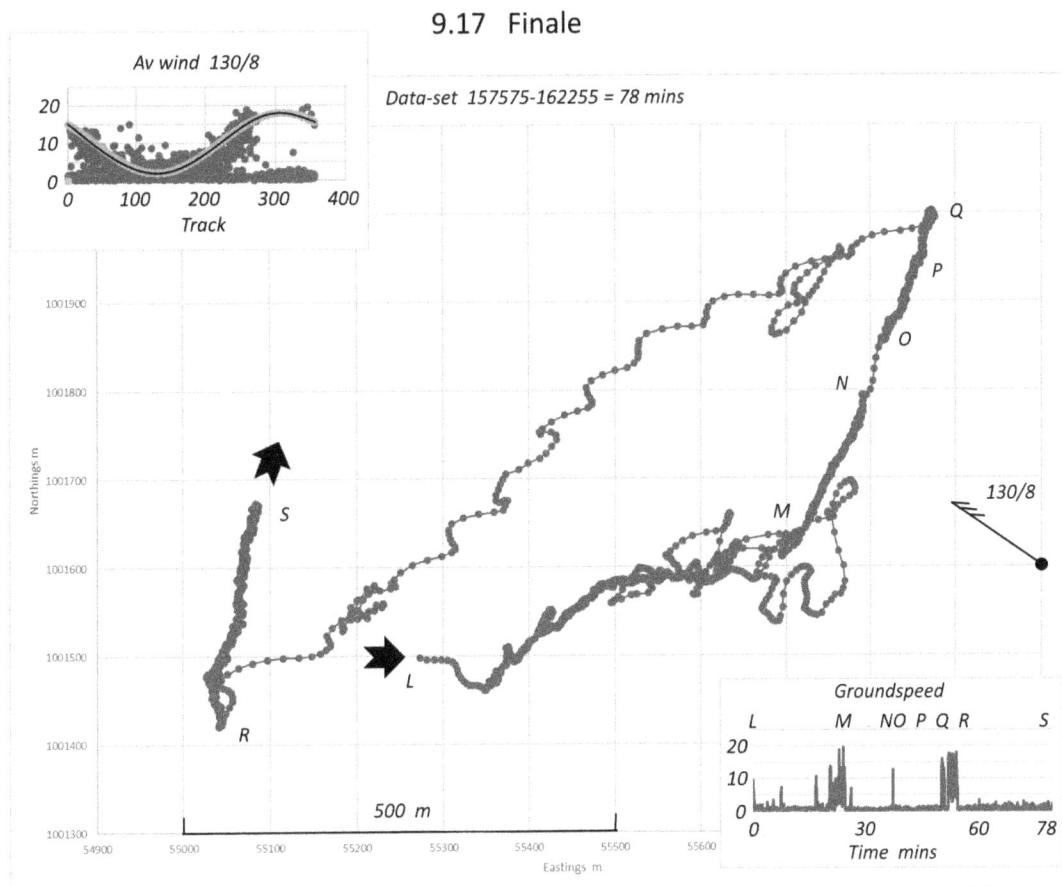

Summary

During his 42 hour journey, it appears that Phil's dynamic soaring technique improved rapidly; so that it might be inferred that he was indeed a juvenile on his first flight.

Phil was airborne for over 88% of the time but, due to the sinuous nature of dynamic soaring, the distance flown was about 75% greater than the straight line distance. It is impossible to know how much flapping was involved. Anecdotally we can estimate there was very little but it appears that when the wind is very light the albatross alternates between flapping and dynamic soaring.

The characteristic sinuous shapes of the tracks show evidence of different dynamic soaring techniques being used to travel crosswind, upwind and downwind. In particular the downwind zig-zag pattern is quite distinctive. In later chapters it will be shown that application of the Windward Turn Theory to these patterns shows that height and speed can be sustained.

There is strong evidence that the albatross foraging technique is to travel crosswind until some trace of prey is detected, presumably through scent trails drifting downwind and then to turn upwind to home-in on the target.

Now we have seen what albatrosses get up to and we understand the theory of flight, we can look at the dynamic soaring manoeuvre in more detail.

Chapter 10

The Windward Turn Theory

Albatross dynamic soaring

The GPS tracking data which were analysed in chapters 8 and 9, give the best picture of what dynamic soaring actually looks like. Even so, to explain what is going on we have to simplify the process and reduce the manoeuvre to a single pair of windward and leeward turns. This chapter will be a narrative description of crosswind dynamic soaring before we get into the mathematics.

In simple terms, dynamic soaring is when the horizontal wind is used to sustain a glider in flight. But dynamic soaring is complicated because there are several interdependent things going on at the same time and we can only explain one thing at a time, so there will always need to be a bit of back and forth in the explanation.

Here are the things we have to understand:

1 - how the groundspeed and airspeed change during upwind and downwind turns

2 - how the aerodynamic forces cause accelerations of airspeed and groundspeed

3 - how the bird's rate of turn causes the headwind component to change and how that affects the airspeed

4 - how the aerodynamic forces are caused by imparting momentum to the air and how the momentum and thus energy is transferred back and forth between the wind and the bird

5 - why the leeward turn is flown as a wing-over

6 - how the wind gradient affects the airspeed

7 - how ground effect works in the windward turn

In Chapter 1, the basic effect of the wind was mentioned; the airspeed of the bird, glider or aircraft will be affected by any *rate of change* of the headwind component, while inertia will resist any change to the actual-speed or groundspeed of the aircraft. That rate of change of the headwind component could be caused by a change in the wind, in other words by turbulence, while the aircraft is essentially flying in equilibrium. But that is not what is happening in dynamic soaring. In dynamic soaring the rate of change of the headwind component is caused primarily by the albatross turning or secondarily by climbing and descending through a wind gradient.

To be clear, the aerodynamic forces acting on the aircraft are not caused by the wind directly. The aerodynamic forces are generated by the relative airflow, caused by the motion of the aircraft through the air; in other words, by airspeed. However, it should be obvious that if an aircraft makes a turn from an upwind heading, through 180 degrees to a downwind heading, the headwind component begins at a maximum, reduces to zero passing the crosswind heading and then increases as a negative headwind, that is a tailwind. Therefore, there has been a rate of change of the headwind component during the turn.

In this chapter we will look at how albatrosses turn this effect into a manoeuvre that enables them to maintain average airspeed and height. In later chapters we will see how albatrosses soar upwind and downwind and how RC glider dynamic soaring works.

The windward turn

The windward turn starts with a tailwind component, turns across the wind and ends with a headwind component. (Figure 10.1 A-B) In the windward turn, the albatross flies a curved path in gliding flight with a small angle of bank, maintaining height close above the surface and flying in ground-effect. Its average course, or track-made-good, after several dynamic soaring manoeuvres, is approximately straight but its heading changes continuously within +/-20 to 30 degrees of the crosswind heading. The difference between its heading and its track is the angle of drift. The bird loses groundspeed and therefore it loses momentum due to an unbalanced horizontal component of the aerodynamic force.

The bird gains airspeed because of the increasing headwind component as it turns. The wind itself need not change; the effect is entirely due to the bird's rate of turn. The albatross flies the windward turn with the least angle of bank which will give it a rate of turn sufficient to maintain airspeed and height, thereby maximising the distance flown by having the greatest radius of turn. Flying close to the surface means that ground-effect improves the efficiency of the turn by reducing drag.

10.1 The Dynamic Soaring manoeuvre

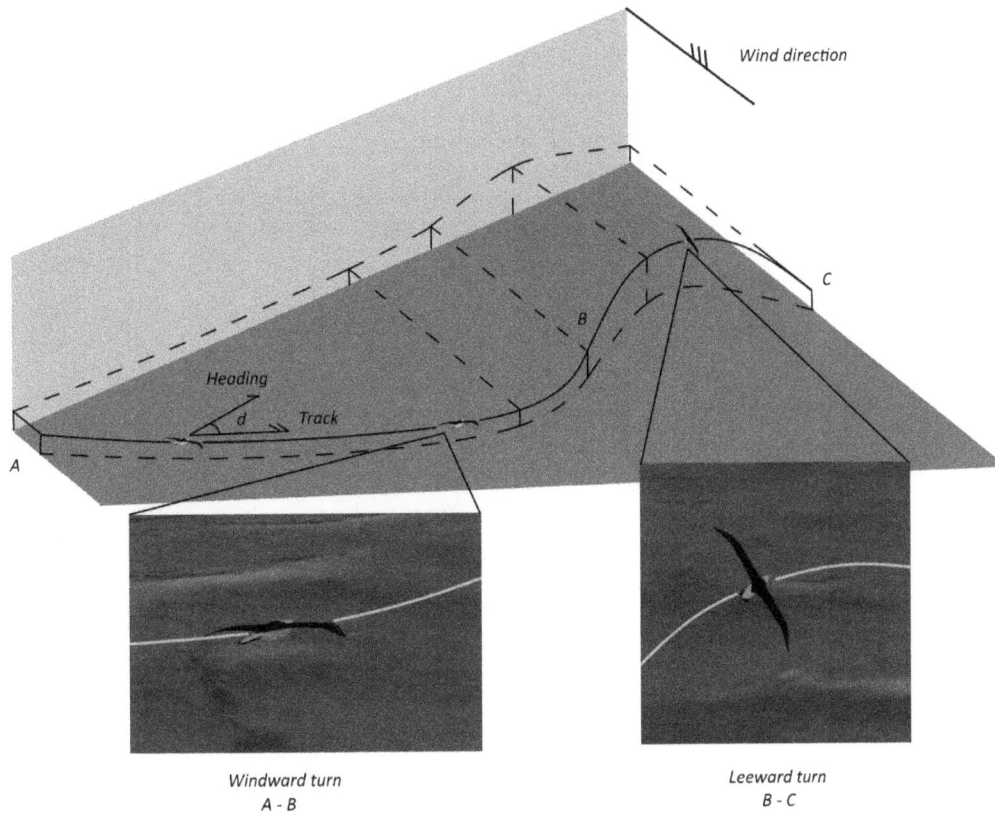

Wind direction

Heading

Track

d

A

B

C

Windward turn
A - B

Leeward turn
B - C

Forces in the windward turn

There are two forces acting on the bird. They are: gravity acting vertically downward and the aerodynamic resultant force comprising the sum of all the small aerodynamic forces acting on the bird by virtue of its airspeed. The resultant is conventionally divided by aerodynamicists into two components; lift and drag. In straight flight, lift and drag lie in a vertical plane.

When the bird banks in a turn, the lift force tilts with the bird's angle of bank and is increased slightly by pitching-up. With a 10-degree angle of bank the lift force is 1.015 times the weight and then has a vertical

10.2 Forces in the windward turn

Straight and level
Lift = weight

10 degrees bank
Lift = 1.015 W

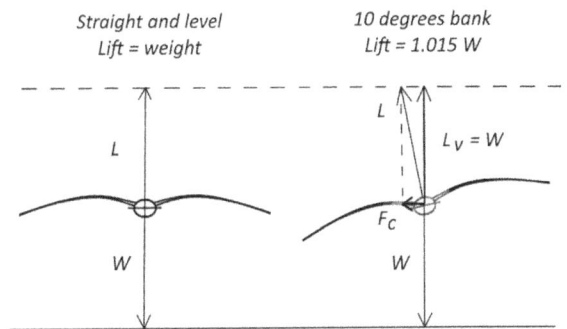

L

W

L

$L_v = W$

F_c

W

and a horizontal component. (Figure 10.2). The vertical component of lift balances weight in level flight and the horizontal component of lift acts as a centripetal force $\mathbf{F_C}$ causing the bird to turn. The centripetal force is normal to the direction of the air-velocity.

10.3 Force components in the windward turn
Plan view Turning left

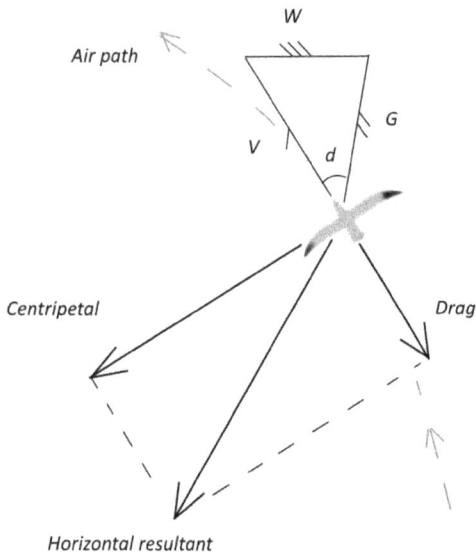

W

Air path

G

V   d

Centripetal   Drag

Horizontal resultant

(a) Forces relative to air-velocity

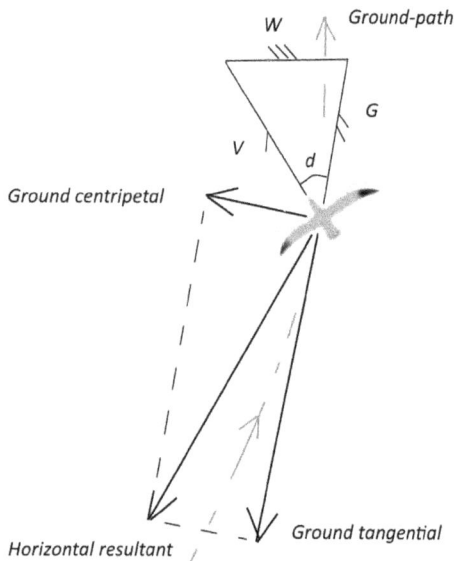

W   Ground-path

G

V   d

Ground centripetal

Horizontal resultant   Ground tangential

(b) Forces relative to ground-velocity

V Air-velocity
G Ground-velocity
W Wind-velocity

In Figure 10.3 the bird is in a windward turn, turning left with the wind from the left. It is a plan view therefore, vertical components are not shown. In a level turn there are now two horizontal components; the drag force and the centripetal force which together form the horizontal-resultant. (10.3a). The horizontal-resultant then forms two more force components orthogonal to the ground-velocity **G**. (10.3b). The ground-centripetal force component is normal to the ground-track and acts as a centripetal force giving curvature to the ground-track. The ground-tangential component is tangential to the ground-track and causes acceleration of the groundspeed. In the windward turn, it is opposite to the ground-track and the loss of momentum is therefore a loss of groundspeed.

The effect of the headwind component

The bird's rate of turn relative to the wind direction causes the headwind component to increase in the windward turn. Airspeed is constant or increases slightly, because the tendency to lose airspeed due to the reducing groundspeed is balanced by the tendency to gain airspeed from the increasing headwind component (a reducing tailwind component has the same effect). The wind-velocity itself need not change, the rate of change of headwind component is entirely due to the bird's rate of turn. While turning close to cross-wind headings, the rate of change of the headwind component is at a maximum even though the actual head/tail-wind component is at a minimum.

In the GPS data in chapter 8 we saw the loss of groundspeed in the windward turn and, after applying an assumed wind (with or without a wind-gradient), we saw the gain of airspeed.

In the windward turn, how do the velocities change?

Figure 10.4 is a plan view of the windward turn showing the triangles of velocity at four points. Note the drift angle d and the wind-angle y. It can be seen that the wind-speed is a large proportion of the birds airspeed and the amount of turn is limited to less than about 60 degrees. During the windward turn there is a large change of wind-angle and a large change of wind-component from tailwind (-H) to headwind (+H). The triangle of velocities itself does not cause the velocities to change, it is simply a way of depicting the three velocities at each moment in time. Those velocity changes are caused by applied forces like lift, drag or gravity.

The bird maintains height and loses momentum due to the aerodynamic forces. The loss of momentum is seen as a loss of groundspeed but not a loss of airspeed. The wind-speed and direction are constant but the wind-angle, the direction of the bird relative to the wind, is changing. Note how the shapes of the triangles are continuously changing due to the bird turning relative to the wind and that the ground-velocity (actual velocity) is reducing as the wind-angle y reduces. Note also that the drift angle d is relatively constant in the range of wind-angles used by the albatross.

10.4 The Windward turn

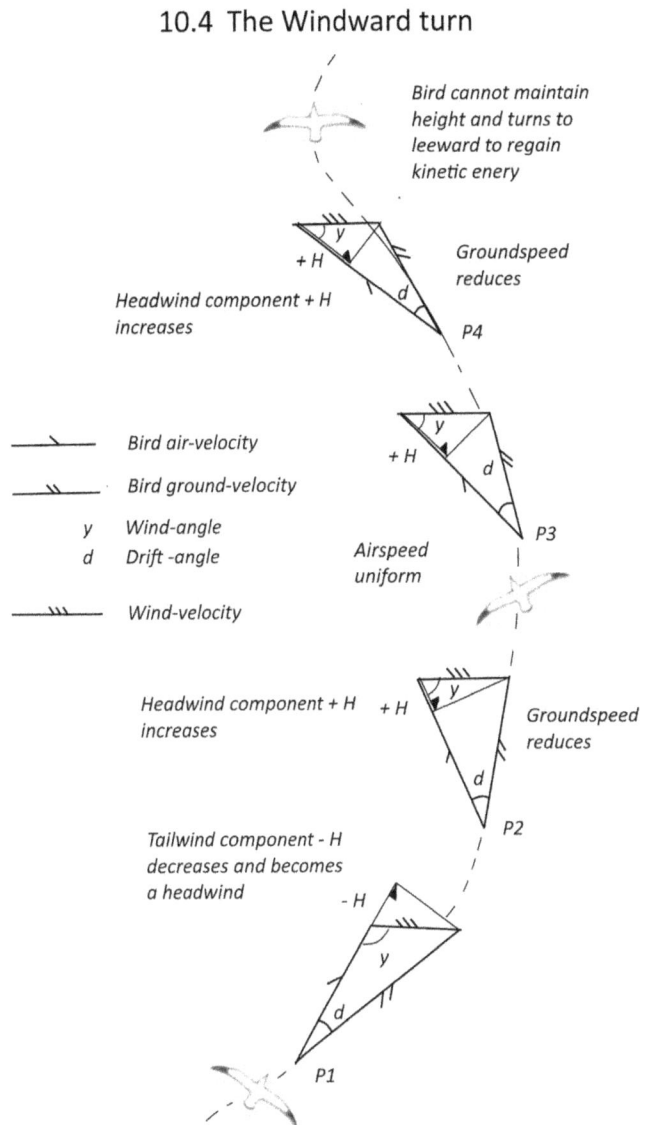

Bird cannot maintain height and turns to leeward to regain kinetic enery

Groundspeed reduces

+H

Headwind component + H increases

P4

+H

P3

Airspeed uniform

Bird air-velocity

Bird ground-velocity

y   Wind-angle

d   Drift -angle

Wind-velocity

Headwind component + H increases   +H

Groundspeed reduces

P2

Tailwind component - H decreases and becomes a headwind   - H

P1

Maintaining airspeed

In dynamic soaring the bird is never in a state of equilibrium, it is always accelerating and of course, there is no thrust. How can the bird, in level flight, maintain its airspeed? In the windward turn the airspeed effectively comprises a ground-speed component and a head/tail-wind component. When the acceleration of the groundspeed component is equal and opposite to the acceleration of the wind component, the acceleration of airspeed is zero. You can see in figure 10.4 that, as the wind-angle y reduces, the tailwind component reduces and then, after the crosswind position, the headwind component increases. At the same time, due to aerodynamic forces, the groundspeed reduces. When these two effects are equal and opposite, the airspeed is constant.

Controlling height and airspeed with rate of turn

The albatross probably judges height above the surface visually and senses airspeed by dynamic pressure through tube-nostrils. The sea surface is constantly rising and falling and therefore the bird must climb and descend to maintain constant height above the surface to take advantage of ground-effect. Climbing and diving to change true height will result in changing airspeed but to maintain airspeed the bird can alter its rate of turn by changing its angle of bank. If airspeed reduces due to the bird gaining height, the bird recovers the airspeed by increasing its rate of turn toward the wind by increasing its angle of bank. If airspeed increases, as for example when descending into a trough, the bird reduces its angle of bank and reduces its rate of turn, reducing its airspeed. Although its airspeed is changing all the time, by minimising airspeed changes it will reduce overall drag losses and improve efficiency and distance gained.

As the windward turn proceeds, the wind-angle reduces, the rate of change of the headwind component inevitably reduces and ultimately the albatross will not be able to maintain its airspeed. Before this happens (Figure 10.4 point P4), the albatross can increase its rate of turn to gain a margin of airspeed and height before reversing the direction of turn into the leeward turn. The penalty here is a small increase in drag, which reduces the total distance flown in the windward turn.

The leeward turn

At the end of the windward turn, the albatross pitches-up and reverses the direction of turn, making the leeward turn as an arched-turn or wing-over at a steep angle of bank (Figure 10.1 B-C). The turn starts with a headwind component and ends with a tailwind component. Despite the large bank-angle, the load-factor is probably only about 1G and therefore requires minimum effort from the bird to keep its wings extended. This is a reasonable assumption, judging by the apparent ease with which the bird maintains is wing posture during the wing-over; although the load-factor experienced by albatrosses during dynamic soaring has never been measured.

The amplitude of the leeward turn is the same as that of the windward turn in order to maintain a constant average angle to the wind and a straight average track on a scale of kilometres. Groundspeed increases during this turn because an aerodynamic force component acts in the direction of the ground-velocity. Airspeed reduces overall due to the decreasing headwind component while there is an exchange of speed and height during the climb and descent.

The leeward turn as a wing-over

In dynamic soaring, the leeward turn is a wing-over or arched turn, which enables a large angle of bank to provide a large horizontal component of lift, without actually increasing the total lift force and without increasing the load-factor and the drag loading (figure 10.5). A wing-over has a partially ballistic trajectory, with the weight of the bird partly supported by a small vertical component of lift. In a 45-degree <u>level</u> turn the load factor will be 1.4G to maintain the vertical force component equal to the weight. However, a 70 degree wing-over can be flown at 1G giving a large horizontal force component but only a small vertical force component. The turn is <u>not</u> level, the bird starts in a climbing turn with momentum and kinetic energy which converts to height and back to speed again during the descent. With a large angle of bank, the horizontal component of lift $\mathbf{F_c}$, added to the drag, creates a horizontal-resultant.

10.5 Forces in a turn and in a wing-over

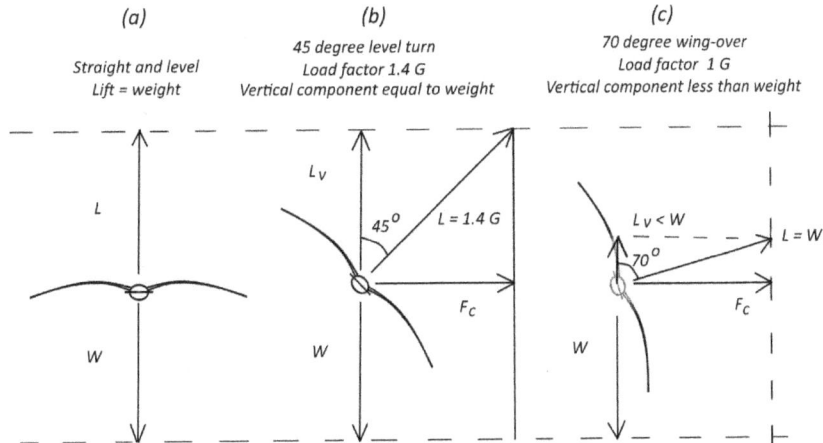

(a)	(b)	(c)
Straight and level	45 degree level turn	70 degree wing-over
Lift = weight	Load factor 1.4 G	Load factor 1 G
	Vertical component equal to weight	Vertical component less than weight

L_V

L $45°$ $L = 1.4 G$ $L_V < W$ $L = W$

$70°$

F_C F_C

W W W

Figure 10.6 is a plan view of the leeward turn as a wing-over to the right, the wind from the left. The large drift-angle is due to the wind being a large proportion of the bird's airspeed. Again, only horizontal force components are shown. The drag force and the centripetal force again form the horizontal resultant (10.6a) but now the ground-tangential force acts in the same direction of the ground-velocity **G**, causing an increase in ground-speed and momentum (10.6b). Thus the bird gains horizontal momentum and kinetic energy, that is groundspeed, without losing potential energy other than a small drag loss during the turn reversals. As this is a wing-over and height is gained and lost, there will be a corresponding loss and gain of airspeed. The wing-over gives the bird time to accelerate, to keep-up with the wind as it turns downwind.

In the leeward turn, (as in the windward turn) the airspeed effectively comprises a ground-speed component and a head/tail-wind component. When the acceleration of the groundspeed component is equal and opposite to the acceleration of the wind component, the acceleration of airspeed is zero. In the leeward turn, the groundspeed is increasing under the effect of the propulsive ground-tangential force component, meanwhile the headwind component reduces under the effect of the increasing wind-angle.

We are used to the idea that drag causes loss of airspeed. However, when turning in a wind, the drag

10.6 Force components in the leeward turn
Plan view turning right

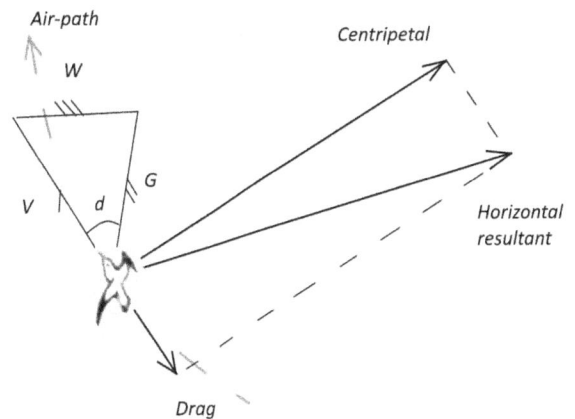

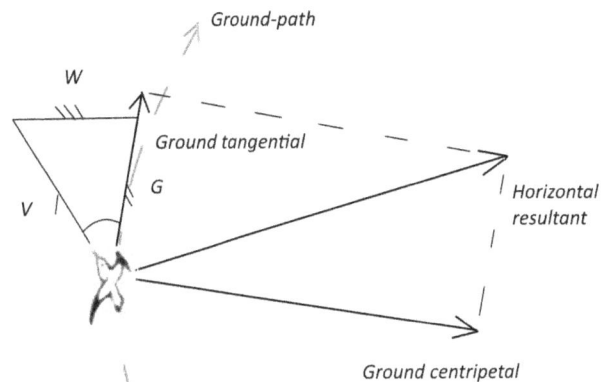

Air-path

W

G

V d

Centripetal

Horizontal resultant

Drag

(a) Forces relative to air-velocity

Ground-path

W

Ground tangential

V G

Horizontal resultant

Ground centripetal

(b) Forces relative to ground-velocity

V   Air-velocity
G   Ground-velocity
W   Wind-velocity

force is only part of the horizontal resultant which causes acceleration of the ground velocity (speed and direction), which then causes acceleration of the groundspeed component which is part of the air-velocity.

In practice, in the range of wind-angles used by the albatrosses, the rates of change of the groundspeed and headwind components are never exactly the same. The magnitude of the rate of change of the headwind component is always slightly greater than that of the groundspeed component. Therefore, the manoeuvre will normally lose a small amount of airspeed in the leeward turn but will make up the loss with a gain in the windward turn. This is due to the different shapes of the windward and leeward turns and the different rates of exchange of momentum.

GPS tracking data clearly shows that there are times when the birds are gaining both groundspeed and height; in other words, they are gaining both kinetic energy and potential energy and not simply exchanging one for the other. This gain of energy can only be achieved by aerodynamic forces and not by wind-gradient effects.

Why does the bird reverse the direction of turn?

At the end of the windward turn, the bird has maintained airspeed at the cost of losing groundspeed. Maintaining airspeed depends upon the ability to match the rate of loss of groundspeed component with the rate of increase of headwind component, which depends upon the rate of change of headwind component relative to the wind-angle. The crosswind position is where the rate of change of headwind with respect to the wind-angle is at a maximum. As the turn proceeds and the wind-angle reduces, the rate of change of headwind-component reduces, whilst drag and the rate of change of ground-speed is approximately constant. The end of the windward turn comes when the bird's ability to maintain airspeed and height diminishes. It must now regain groundspeed and it can only do this by turning downwind.

At the end of the leeward turn the bird has restored its groundspeed and momentum but has run out of height and airspeed. It must again reverse its direction of turn and start another windward turn.

The ability to maintain height depends on the rate of change of head/tailwind component, which is greatest in the middle part of the windward turn and diminishes when turning to within about 50 degrees of the wind. For this reason, the albatrosses do not do dynamic soaring in circles or in alternating 180 degree turns as assumed by the wind-gradient theory.

Transfer of Energy and Momentum

How exactly is energy exchanged between the wind and the bird? Fundamentally, the process is a transfer of momentum, the same principle as colliding pool balls. The momentum of the wind depends on wind-speed and that is measured relative to the ground. This does not make the ground a privileged frame of reference. It is simply a convenient, common frame of reference relative to which to measure the wind-velocity and the bird's velocity and hence its acceleration. When the wind gains speed and momentum, it is the air which is accelerating, not the ground. The bird's actual velocity is, of course, the vector sum of its air-velocity and the wind-velocity.

Momentum is mass times speed. A rate of change of momentum is mass times acceleration. According to Newtonian mechanics: Force equals mass times acceleration. Thus, the lift and drag forces are the equal and opposite reactions to the rate of change of momentum given to the air by the motion of the wing through the air. When a bird or an aircraft banks in a turn, the lift force tilts with the bird and the horizontal component of lift acts as a centripetal force which makes the bird turn. This horizontal centripetal force means that horizontal momentum is given to the air, opposite to the centripetal direction. (Figure 10.7).

10.7 Transfer of momentum in leeward and windward turns

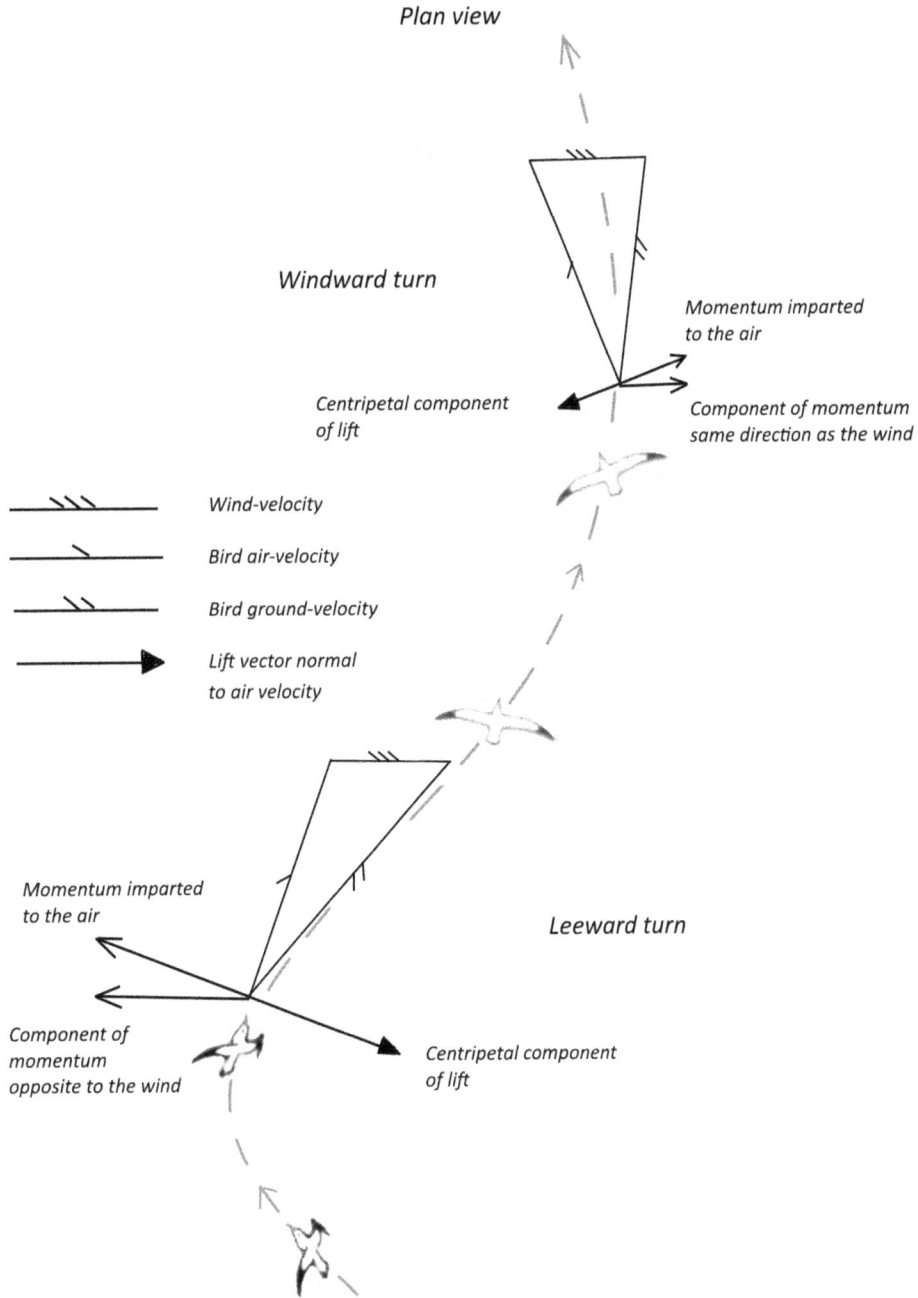

Plan view

Windward turn

Momentum imparted to the air

Centripetal component of lift

Component of momentum same direction as the wind

———⧸⧸⧸——— *Wind-velocity*

———⟍——— *Bird air-velocity*

———⟍⟍——— *Bird ground-velocity*

———▶ *Lift vector normal to air velocity*

Momentum imparted to the air

Leeward turn

Component of momentum opposite to the wind

Centripetal component of lift

In the windward turn, a component of this horizontal momentum is in the *same* direction as the wind and gives momentum and energy to the wind. As the albatross loses momentum, so the wind gains momentum. The windward turn is flown with a small angle of bank and a small horizontal component of lift and therefore the turn is long and the exchange of momentum is at a slow rate.

In the leeward turn, the bird's change of direction is again caused by a centripetal force, a horizontal component of the lift force, the result of the wing imparting horizontal momentum to the air. A component of that momentum is now *opposite* to the wind direction, therefore the momentum of the wind is reduced while the momentum of the bird is increased. The forces are relatively large due to the large angle of bank and therefore, the rate of change of momentum is relatively quick but the length of the leeward turn is relatively short. The momentum in both turns and for both the bird and the air is measured relative to the same ground frame of reference.

The bird maintains average speed and height during successive windward and leeward turns therefore its average momentum is constant but its speed changes continuously during each turn. For one unit-mass of bird, the change of momentum with speed is about the same in each turn. However, as the bird flies, it encounters successive <u>different</u> unit masses of air. The leeward turn is shorter than the windward turn and fewer unit masses of air are each given greater acceleration; compared with the windward turn which is longer and in which a greater number of unit masses of air are each given less acceleration. So, per unit mass of air, the wind loses more speed in the leeward turn than it gets back in the windward turn.

Kinetic energy is proportional to mass times speed squared. The bird's average kinetic energy is constant but with its constantly changing speed it is constantly exchanging both momentum and kinetic energy with the wind. In the windward turn the kinetic energy of the bird decreases and that of the wind is increased; in the leeward turn the kinetic energy of the bird increases and the kinetic energy of the wind is reduced. Overall, the kinetic energy of the wind is reduced after the passage of the bird. The difference in energy due to the reduction of wind-speed, is equivalent to the bird's drag losses. In effect, wind-speed energy has been converted into air-turbulence energy in the wake of the bird. Ultimately that turbulence energy is converted to heat energy at the molecular level and dissipated throughout the air.

In this way, the albatrosses are able to bias the exchange of momentum and kinetic energy in their favour and, provided they gain slightly more airspeed and height in the windward turn than they lose in the leeward turn, then they can continue soaring indefinitely as long as the wind blows.

The effect of the wind gradient

Wind gradients occur as a result of friction between the air and the ground, independent of any bird activity. The result is a reduction of the average wind speed and therefore less wind energy available for the bird. Climbing upwind and descending downwind through a wind gradient will tend to increase airspeed but only if actual speed is maintained. However, in a straight, upwind climb, actual speed must reduce due to gaining height and doing work against gravity. In this case, the effect of the wind gradient will be to reduce the airspeed loss but not to increase the airspeed. In a downwind descent through a wind gradient, actual speed will only be maintained by an ever steeper angle of descent as the drag load increases with the increased airspeed. In practice, what will happen is that there will be a small increase in airspeed which will result in a small increase in drag-force which will cause the actual-speed to reduce. The result is that actual-speed reduces continuously until the descent through the wind gradient is complete, at which point the small excess airspeed is dissipated and a new state of equilibrium is established.

In dynamic soaring, the bird's height and ground-speed can both increase at the same time, as seen in the GPS data, but this is due to the aerodynamic force component, not the wind gradient. The effect of the wind gradient may be to improve the efficiency of the leeward turn during the climb and descent by reducing the loss of airspeed overall. This will be a relatively small effect because the birds fly mainly close to cross-wind headings where the head-wind component is at a minimum.

Ground effect

During the windward turn, the bird flies close to the surface to gain advantage of the ground-effect. This is where the lift-induced drag of an aircraft is reduced by the close proximity of the surface.

The lift and drag forces are the equal and opposite reactions to the downward and forward momentum given to the air by the wing's down-wash. The rate of vertical momentum given to the air is always equal to the weight in straight flight; the rate of horizontal, forward momentum given to the air is equal to the drag-force. However, when the wing is within about half a wingspan of the surface, the forward momentum given to the air is reduced and the wing produces the same amount of lift with less of a drag penalty. Ground-effect does not add energy but improves the efficiency of flight by reducing the drag-load. It increases the distance flown and provides the incentive for the bird to remain close to the surface.

The nature of the wind

In the Windward Turn Theory, crosswind dynamic soaring is possible in a uniform wind, which is to say a wind-gradient is not essential. However, a uniform wind is not the same as a box of air moving with uniform velocity. If that were the case, then you could take that box of air to anywhere in the universe and it would continue moving with uniform velocity unless acted upon by an external force, according to Newtons Laws of Motion.

In practice, wind speed and direction are measured relative to the surface of the Earth; wind-velocity on planet Earth is caused by temperature and pressure variations within the atmosphere. Wind energy is derived from the energy of the Sun causing surface heating; that heat then being transferred to the air by conduction and convection. The temperature difference between one place and another, causes a pressure gradient which then causes the wind to move over the curved surface of the Earth, deflected by Coriolis effect.

For our purposes, describing the effect of turning in a wind, the wind is taken to be uniform; that simply means that it has a particular velocity at a particular place and time when encountered by the albatross. It also means that, in a mathematical model of dynamic soaring, there need not be any wind gradient, although a wind gradient effect can be added into the calculations optionally. When the albatross passes and aerodynamic forces are generated, the air becomes accelerated and the wind-velocity at that point increases or decreases a little bit. In that sense the wind is not uniform. Overall, the wind loses some speed but the energy returns to the air in the form of turbulence in the wake of the bird.

Further development of the Windward Turn Theory shows that upwind dynamic soaring is possible and in this case a wind-gradient is necessary to promote the rate of change of the headwind component and the drift angle.

(Coriolis effect is when a moving air-mass is deflected to the right in the Northern hemisphere and to the left in the Southern hemisphere. This is because, while the Earth rotates with a particular angular velocity, the speed of the surface moving Eastwards is greatest at the Equator and is slower at higher latitudes as the surface is closer to the axis of rotation. An air-mass which is stationary relative to the surface of the Earth is actually moving with the same velocity as the surface at the particular latitude. When the air-mass moves to a higher or lower latitude, inertia preserves its initial Eastward velocity while the new surface is moving either slower or faster; the result is that the air-mass turns to the right in the Northern hemisphere and to the left in the South).

Airspeed sense

The need to maintain or slightly increase airspeed in the windward turn and the need to sense when airspeed is changing, is critical to dynamic soaring and controls the bird's management of the manoeuvre. The points at which the turn-reversals are made, from the windward to leeward turn and vice-versa, depend upon the bird's ability to sense when the relationship between airspeed (dynamic pressure) and height (visual) is changing.

This may explain why birds that do dynamic soaring for a living, the albatrosses and petrels, are found to have nostrils of the tube-nose sort. They may make these birds particularly sensitive to dynamic pressure and therefore to *rate of change* of airspeed. I suspect that, in other forms of bird flight, angle of attack and groundspeed are more important than pure airspeed.

Tube nostrils are similar to the pitot tubes found on aircraft. Pitot tubes are the forward facing vents which feed pitot pressure to an aneroid capsule in the airspeed indicator (ASI). Pitot pressure is the sum of static (ambient) pressure and dynamic or impact pressure, caused by the relative airflow. Static pressure from a separate vent, is led to the sealed case of the instrument. The expansion of the capsule in the case is then proportional to dynamic pressure which is the difference between pitot pressure and static pressure. The expansion of the capsule is transferred by levers and gears to the instrument needle and the ASI then reads indicated airspeed. However, indicated airspeed will depend on air density which varies with altitude and temperature, so that the instrument is only correct at sea level, at a standard temperature. True airspeed can be calculated given altitude and temperature data. The picture shows a giant petrel and its tube nostrils

Measuring true height requires an altimeter with a sealed aneroid capsule. Emulating such a device in nature with a sealed air-tight organ would not be practical; typically, vertebrate inner ears are vented to the throat through the Eustachian tubes. Most birds have sideways facing nostrils which appear to have evolved to minimise the effect of dynamic pressure and make it easier for the bird to breathe, especially when diving at high speed. Whether the nasal airways of soaring birds are sensitive to static pressure is not known. Static pressure varies with height but also varies from day to day and from place to place, so that height is not exactly indicated by static pressure. It may be useful to a soaring bird to sense changes of air-pressure with height in order to detect thermals. Knowing whether it is gaining or losing height would be more useful to a bird than knowing its exact height. They would not be measuring height by sensing absolute pressure but rather they would be sensing rate of climb and descent by sensing rate of change of air pressure. The alternative would be to detect height changes visually which would be increasingly difficult at great heights.

It would make some sense if all birds have the ability to sense rate of change of air pressure which is then adapted to the particular needs of either dynamic or regular soaring.

This chapter has been about describing how the crosswind dynamic soaring manoeuvre appears to work with a uniform wind without any wind-gradient. The next three chapters are about the basic effect of the wind and the development of a mathematical model of dynamic soaring which simulates how the albatrosses and model gliders fly.

Chapter 11

The effect of the wind during acceleration

If you ask most pilots if the wind affects airspeed they will reply no, the wind only affects groundspeed. However, they are automatically thinking of what happens in equilibrium and in navigation where ground-velocity is the vector sum of air-velocity and wind-velocity. But equilibrium is simply a special case of zero acceleration of the aircraft where the wind does indeed have a zero effect on airspeed. The general case is what happens during various degrees of acceleration, that is to say rates of change of speed and direction. When the aircraft acceleration is non-zero, the effect of the wind on airspeed is also non-zero. The simplest way to visualise these effects is to consider the effect of an unbalanced drag force on a glider in straight flight.

The effect of the wind on a glider during straight acceleration

When a glider maintains height on a constant heading, there is no thrust and consequently the unbalanced drag-force causes the air-speed to reduce. For simplicity we can assume, for the moment, that drag is constant even as airspeed reduces. See figure 11.1 which shows a plan view of two successive positions of a glider in straight flight at constant height. Only horizontal components are shown and so lift and the vertical component of lift are not shown. The wind blows from the left at constant velocity.

Force components

Force $\mathbf{F_D}$ is the unbalanced drag force, the air-tangential force, opposite to the glider's air-velocity. The glider is not turning relative to the air, its wings are level, the air-centripetal acceleration is zero and is therefore not shown.

The effect of the wind is to cause an angle of drift **d** between the air-velocity and the ground-velocity. Also, in this case, the drag force is in-line with the air-velocity vector but is at an angle, the same drift angle **d**, to the ground-velocity vector. The aircraft 'knows' its drift angle because the tangential *accelerations* of airspeed and ground-speed are different and are separated by the angle of drift.

The total force and total acceleration are the same in both air and ground frames of reference. However, relative to the ground-velocity, drag force $\mathbf{F_D}$ can be divided into ground-tangential component $\mathbf{F_{GT}}$, opposite to the ground-velocity and ground-centripetal component $\mathbf{F_{GC}}$ normal to the ground-velocity. These components provide tangential and centripetal accelerations of the ground-velocity.

$\mathbf{F_D}$ is the vector sum of component $\mathbf{F_{GT}}$ and $\mathbf{F_{GC}}$

$$\overrightarrow{\mathbf{F_D} = \mathbf{F_{GT}} + \mathbf{F_{GC}}}$$

11.1 Effect of the wind on a glider in straight flight with an unbalanced drag force

Plan view of two successive positions

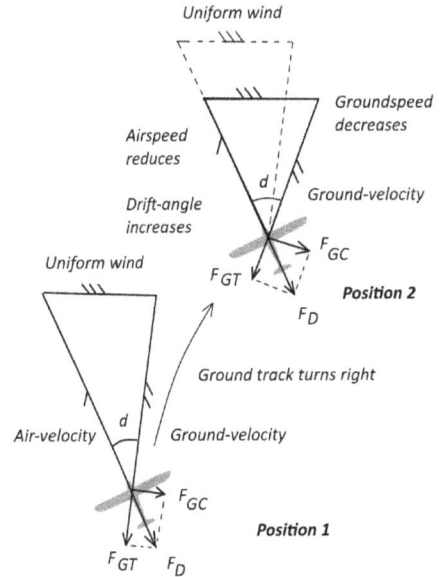

Relative to the air, the glider is flying straight
As the airspeed reduces, the drift-angle increases resulting in a curved path relative to the ground

F_D Drag force
F_{GT} Ground tangential force
F_{GC} Ground centripetal force

As the airspeed reduces, so the groundspeed reduces due to ground-tangential force $\mathbf{F_{GT}}$, the drift angle **d** increases and the direction of the ground-velocity turns to the right. The path over the ground is a curve even though the path through the air is straight. That curve is caused by a ground-centripetal force component $\mathbf{F_{GC}}$ even though the air-centripetal force is zero.

$\mathbf{F_{GT}}$ is less than $\mathbf{F_D}$ and therefore, in this case, the rate of ground-tangential acceleration is less than the rate of air-acceleration. This means that, although the airspeed can in theory reduce to zero, the groundspeed can only reduce to the value of the wind-speed and the direction of the ground-velocity will then be parallel with the wind. Clearly, this abnormal situation can only last for a few seconds before the glider runs out of airspeed and stalls.

This example illustrates how the effect of a single force is different in the two frames of reference, the air and the ground. This is because of the uniform relative velocity of the two frames and the angle between the applied force and the ground-velocity vector. This effect only occurs because the aircraft is accelerating in both frames of reference.

Speed and acceleration components

During acceleration, airspeed **V** can be thought of as having two components: a headwind component **H** and a groundspeed component **K**. Airspeed **V** is the sum of components **H** and **K** (either **H** or **K** could have a zero value). See figure 11.2 which shows the same situation as figure 11.1 but with some different information.

$$V = H + K$$

The headwind component **H** is equal to the product of the wind-speed and the cosine of the wind-angle. The groundspeed component **K** is the product of the groundspeed and the cosine of the drift-angle. The drift angle is the angle between the aircraft heading and its track over the ground.

H = W.cos y   **W** is the wind speed   **y** is the wind-angle
K = G.cos d   **G** is groundspeed   **d** is drift angle

The acceleration of airspeed **V** is the sum of the rates of change of **H** and **K**. Either rate of change could have a zero value in which case the acceleration of **V** will be the same as the non-zero value.

$$R_V = R_H + R_K$$

R_V is the rate of change of **V**
R_H is the rate of change of **H**
R_K is the rate of change of **K**

Straight flight, is a case when there is an unbalanced drag force F_D. Airspeed **V** reduces but the headwind component **H** is constant because the aircraft is not turning and the wind-angle **y** is constant. In other words, when the rate of change of the headwind component R_H is zero, it is the component **K** which is reducing at the rate R_K and therefore the rate of change R_V of the airspeed **V** is the same as the rate of change R_K of the groundspeed component **K**.

$$R_V = R_H + (-R_K)$$

When $R_H = 0$ then $R_V = -R_K$

Summary

This example clearly shows that when there is a uniform wind and an angle of drift, the total force and total acceleration are the same in both air and ground frames of reference but the tangential components are different in each frame of reference and the centripetal components are different in each frame of reference, making a total of four acceleration components. When the aircraft is under acceleration, there can be a situation where one of the four acceleration components is zero while the others have a non-zero value. Specifically, in this case, the air-centripetal

11.2 The effect of the wind during straight acceleration

Plan view of two successive positions

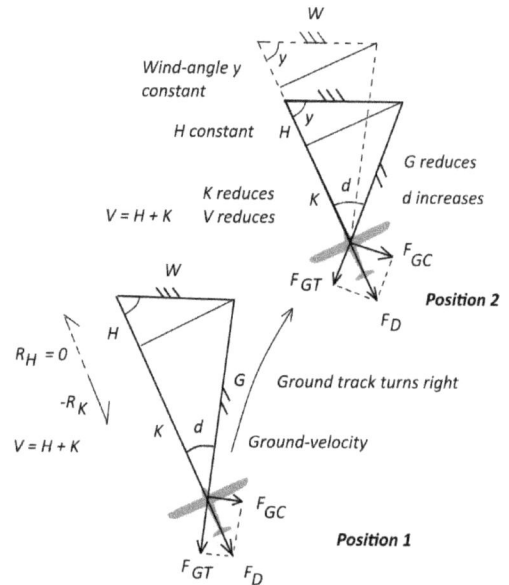

$$V = H + K$$
$$R_V = R_H + (-R_K)$$
$$R_H = 0 \quad \text{therefore}$$
$$R_V = -R_K$$

V Air-velocity	F_D Drag force
G Ground-velocity	F_{GT} Ground tangential force
W Wind-velocity (uniform)	F_{GC} Ground centripetal force

H Headwind component
K Groundspeed component
R_H Rate of change of H
R_K Rate of change of K
R_V Rate of change of V

force is zero and the aircraft flies straight in the air. Meanwhile the ground-centripetal force is non-zero and the aircraft has a curved path over the ground.

In the next chapter things get more complicated. What happens when the acceleration is a turn relative to the wind-direction at constant airspeed?

Chapter 12

The effect of the wind on an aircraft in a turn

In previous chapters we have seen that, for an aircraft in level flight, airspeed is controlled by a balance of thrust and drag or, in the case of a glider, a component of weight balances drag at a constant rate of descent. We are now going to learn that in the general case of non-zero acceleration, particularly when the aircraft is turning, that there are additional effects caused by the wind which have to be taken into account to explain the energy gains and losses during dynamic soaring and general flight. This is additional to the slight loss of airspeed caused by the increased load factor in a turn.

In this chapter, to simplify things, we will consider the effect of the wind on a powered aircraft in level, turning flight where there is no unbalanced drag force and we can just look at the effect of the centripetal force.

We will see that, in a uniform wind, the airspeed of an aircraft which is turning is subject to a cyclical variation in which the airspeed decreases slightly during the downwind turn, when the groundspeed is increasing and increases slightly during the upwind turn, when the groundspeed is decreasing.

These effects create the conditions which contribute to stall-spin accidents during aircraft downwind turns but also create the circumstances in which albatrosses and other seabirds are able to maintain average height and airspeed during dynamic soaring.

Forces in the Windward Turn

Throughout the windward turn, groundspeed reduces. See figure 12.1 which shows two diagrams of the same position. A powered aircraft is turning left and maintaining height in a windward turn, the wind blows from the left. Lift and its vertical component are not shown. Thrust and drag are equal and opposite, there is no unbalanced drag force and therefore force $\mathbf{F_D}$ is not shown. This time, the air-centripetal force $\mathbf{F_C}$ is the unbalanced force. $\mathbf{F_C}$ is the horizontal component of lift acting as the air-centripetal force normal to the air-tangential velocity $\mathbf{V}$.

12.1 Force components relative to air and ground velocities for a powered aircraft in a windward turn

Two plan views of the same position

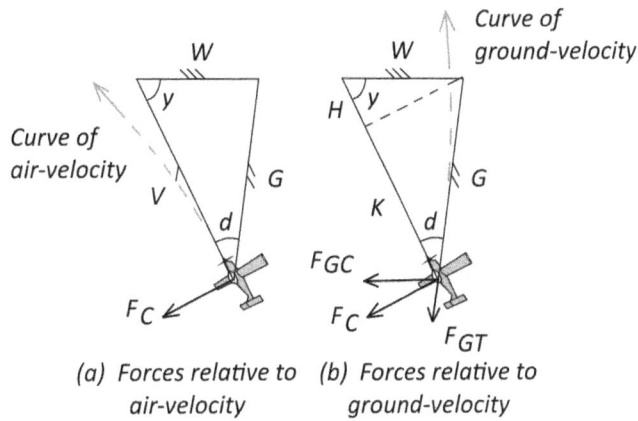

(a) Forces relative to air-velocity   (b) Forces relative to ground-velocity

$$V = H + K$$

V Air-velocity
G Ground-velocity
W Wind-velocity
d Drift-angle
y Wind-angle
H Headwind component
K Groundspeed component

Thrust and drag are equal and opposite and are not shown
F_C Air centripetal force (depends on angle of bank)
F_{GT} Ground tangential force (retarding)
F_{GC} Ground centripetal force

When an aircraft turns at constant airspeed, it is obvious that the groundspeed changes – 'because of the wind'. However, that change of groundspeed is not caused directly by the triangle of velocities, it is an acceleration of a mass which requires a force, according to Newton's laws of motion. What force causes that acceleration?

The answer is that relative to the ground-velocity, force $\mathbf{F_C}$ resolves into ground-tangential force $\mathbf{F_{GT}}$, which causes the groundspeed to reduce in the windward turn and ground-centripetal force $\mathbf{F_{GC}}$, which provides the curvature of the ground track. These force components have a simple trigonometrical relationship depending on the drift angle **d**:

$$\mathbf{F_{GT} = F_C .sin\ d}$$
$$\mathbf{F_{GC} = F_C .cos\ d}$$

Once again, the total force and total acceleration is the same in both air and ground frames of reference but this time $\mathbf{F_C}$ is the vector sum of $\mathbf{F_{GT}}$ and $\mathbf{F_{GC}}$:

$$\overrightarrow{}$$
$$\mathbf{F_C = F_{GT} + F_{GC}}$$

Clearly the two centripetal components are different; $\mathbf{F_{GC}}$ is slightly less than $\mathbf{F_C}$, so the curvature of the ground track is less than that relative to the air:

$$\mathbf{F_C > F_{GC}}$$

Considering the tangential components, there is no unbalanced drag force, therefore $\mathbf{F_D}$ is not shown, while $\mathbf{F_{GT}}$ is non-zero. Compare with the case in fig 11.1 (chapter 11) where $\mathbf{F_C}$ is zero (not shown) while $\mathbf{F_D}$ is non-zero.

Velocity components in the Windward Turn

As in the last chapter, airspeed can be thought of as being the sum of two components **H** and **K**.

$$\mathbf{V = H + K}$$

The components **H** and **K** again have a simple trigonometrical relationship to the wind-speed and ground speed respectively:

$\mathbf{H = W .cos\ y}$ **W** is wind speed   **y** is wind-angle
$\mathbf{K = G .cos\ d}$ **G** is groundspeed   **d** is drift angle

When an aircraft turns relative to a uniform wind, there is a continuous change of the magnitude of the headwind component **H** parallel to the aircraft longitudinal axis. (Figure 12.1 again). This is caused by the aircraft turning and the changing wind-angle, even though the actual wind is not changing. For our purposes the aircraft heading is measured relative to the wind-direction and is called the wind-angle **y**.

However, note that the aircraft does not 'know' it's groundspeed or the wind-speed or the wind-angle. These are values which appear in the mathematical model but the aircraft only 'knows' its airspeed, which produces aerodynamic forces, and the effect of those forces in terms of tangential and centripetal accelerations. It 'knows' its drift angle because of the acceleration of air-velocity and ground-velocity in two different directions, separated by the angle of drift.

In this case of centripetal acceleration in a turn, both components H and K are changing but under different influences. Component H changes when the aircraft is turning because the wind-angle y is changing. That is to say that there is a different value of H for each value of **y**. Component K is changing because the groundspeed is changing as we saw in chapter 11. Acceleration of airspeed is the sum of the rate of change of the headwind component H and the rate of change of the groundspeed component K:

$$R_V = R_H + R_K$$

R_V is the rate of change of V
R_K is the rate of change of K
R_H is the rate of change of H

The rate of change R_H (m/s per second) is the product of the rate of change of headwind component per degree of wind-angle (m/s per degree, depending on the wind-angle) and the rate of turn of the aircraft (degrees per second).

The rate of change R_K depends on the rate of change of groundspeed and on the drift angle. The rate of change of groundspeed is caused by force component F_{GT} acting tangentially to the ground-velocity.

When turning to windward, the wind-angle **y** is decreasing and therefore headwind component H is increasing $(+R_H)$. Groundspeed G is decreasing due to force F_{GT} and therefore component K is decreasing $(-R_K)$. The acceleration of airspeed (R_V) is the sum of the rates of change of H and K. (see figure 12.2 which shows two successive positions).

$$R_V = +R_H + (- R_K)$$

Airspeed, in a turn in a wind, is therefore affected not only by a balance of forces (thrust and drag) but also by a balance of accelerations. **If** the two rates of change are equal and opposite, then the rate of change of the airspeed is zero and airspeed is constant.

*However, when we calculate the values of R_H and R_K we will find they are **not** exactly equal and opposite and therefore there is always a slight variation of airspeed.*

12.2 Effect of changing wind-angle in the windward turn

Plan view of two successive positions

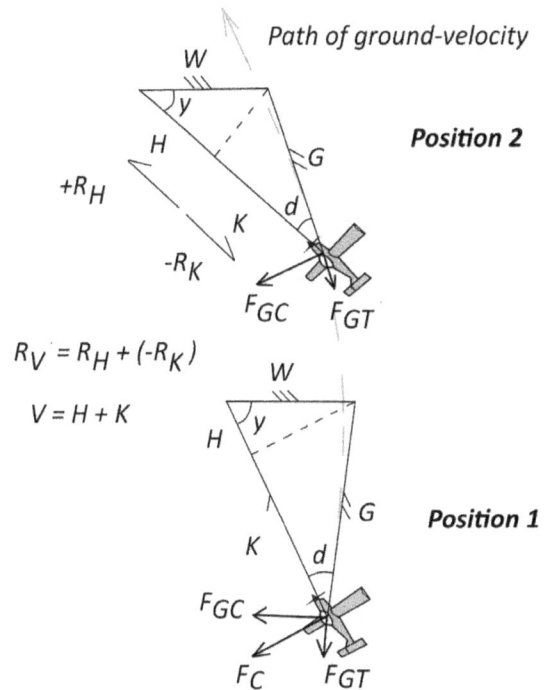

$$R_V = R_H + (-R_K)$$
$$V = H + K$$

As the aircraft turns toward the wind wind-angle y reduces while headwind component H increases
Force F_{GT} causes groundspeed G to reduce

Turning to leeward

Now consider the same powered aircraft in a leeward or downwind turn, during which groundspeed increases. See figure 12.3 below, showing two views of the same position. The aircraft is in a leeward turn banked to the right, with the wind from the left. Thrust equals drag so there is no unbalanced drag force but with the same incremental loss of airspeed due to increased load factor. $\mathbf{F_C}$ is the horizontal component of lift acting as the air-centripetal force. $\mathbf{F_C}$ resolves into $\mathbf{F_{GC}}$, the ground centripetal force and $\mathbf{F_{GT}}$, the ground tangential force which is now <u>propulsive</u> and makes the ground speed $\mathbf{G}$ increase. This is because the angle of bank is now on the *same* side as the angle of drift, in this case turning to the right and drifting to the right.

12.3 Force components for a powered aircraft in a leeward turn

Two views of the same position

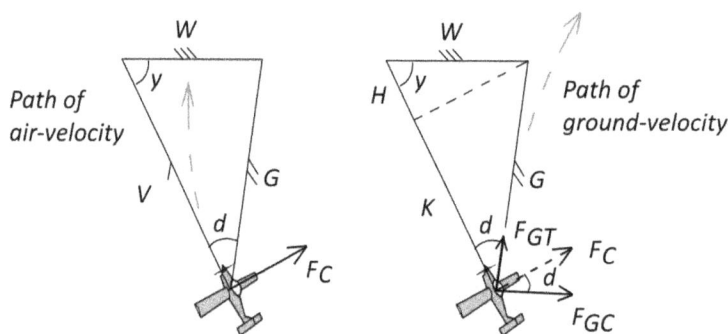

(a) Forces relative to air-velocity

(b) Forces relative to ground-velocity

$$V = H + K$$

Thrust is equal and opposite to drag and is not shown
F_C *is the air centripetal force*
F_{GT} *is the ground tangential force (propulsive)*
F_{GC} *is the ground centripetal force*

Component $\mathbf{K}$ increases as groundspeed $\mathbf{G}$ increases whilst component $\mathbf{H}$ reduces as the wind angle $\mathbf{y}$ increases. (See figure 12.4 showing two successive positions). Once again, it will be shown that $\mathbf{R_K}$ and $\mathbf{R_H}$ are <u>not exactly</u> equal and opposite and there is a slight reduction of airspeed in the leeward turn.

$$\mathbf{R_V} = \mathbf{R_K} + (-\mathbf{R_H})$$

The difference between the windward and leeward turns

During the windward turn, the airspeed increases due to $+R_H$ being slightly greater than $-R_K$. As the airspeed increases, so the drag increases and the rate of decrease of the groundspeed component $-R_K$ increases in magnitude. However, as $+R_H$ is caused by a constant rate of turn and $-R_K$ is caused by a tangential force acting on a mass, the effect of inertia is to resist a change to $-R_K$ but not to affect $+R_H$. The result is that the effect of $+R_H$ is slightly greater than the effect of $-R_K$, leaving an incremental increase in airspeed and a slightly increased rate of loss of groundspeed. This process increases to a maximum rate at the cross-wind position and diminishes at the upwind and downwind positions.

The opposite effect occurs in the leeward turn and the airspeed tends to decrease due to the effect of $-R_H$ being slightly greater than the effect of $+R_K$. In both cases the R_H term is slightly greater in magnitude than the R_K term. In the absence of an unbalanced drag force, these effects are caused only by the centripetal force F_C.

12.4 Effect of changing wind-angle for a powered aircraft in a leeward turn

Plan views of two successive positions

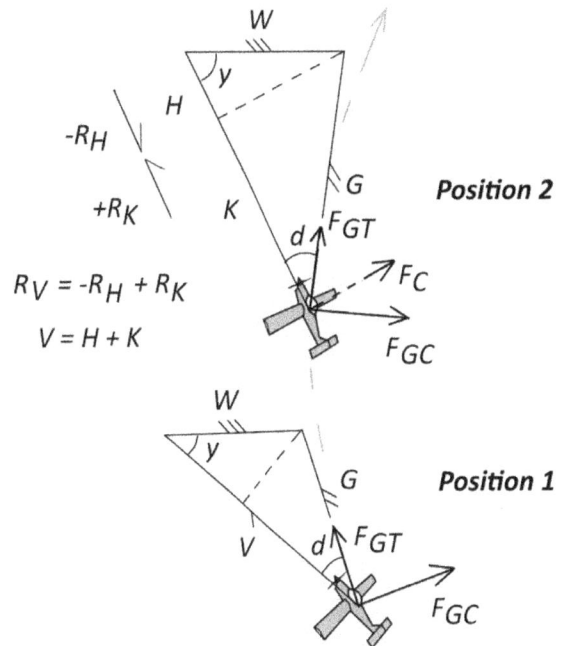

As wind-angle y increases, headwind component H reduces while ground tangential force F_{GT} makes groundspeed G increase and so K increases

*By plotting the accelerations R_H and R_K against the wind-angle during a full 360-degree circle, it can be shown that these two rates of change are **not** exactly equal and opposite and therefore there is **always** a slight rate of change of airspeed. There is a slight increase of airspeed in the windward turn and a slight loss of airspeed in the leeward turn.*

In both turns a simple feedback mechanism responds to the changing headwind component. At a constant angle of bank, the rate of turn and load-factor are constant, whilst the headwind component changes due to the changing wind-angle. The changing airspeed causes a changing drag load which causes a tangential acceleration opposite to the change of airspeed which caused it. Thus the actual change of airspeed is very small. The gain of airspeed in the windward turn is approximately balanced by the loss of airspeed in the leeward turn, so it is easily overlooked in normal aircraft operations. It becomes more significant when the wind is a large proportion of the airspeed and there is a large angle of drift.

Demonstrating the rate of change of airspeed

We can demonstrate these effects by modelling a full circle flown by a powered aircraft. We can derive formulae for R_H and R_K and plot them against wind-angle over a full, 360 degree circle at say 10 degree intervals with a uniform wind. At each wind-angle we calculate groundspeed and drift angle, R_H and R_K and R_V. The time interval is the angle interval divided by the rate of turn R_Y which depends on the centripetal component F_C. Using the time interval and the rate R_V the new V is calculated and carried forward to the next wind-angle. See Appendix 1 for the calculations.

We can now apply some starting values: let's say wind-speed 10m/s, angle of bank 30 degrees and airspeed is initially 50m/s. See figure 12.5 which plots the *rates of change* of headwind component R_H, groundspeed component R_K and airspeed R_V against wind-angle during a 360 degree circle. The graphs confirm our reasoning so far; the rates of change of the headwind component and groundspeed component are indeed greatest on the crosswind headings (90 and 270) and are approximately equal and opposite. However, the *acceleration* of airspeed (the light grey line - the sum of the two other lines) is <u>not</u> exactly zero, being slightly negative during the downwind turn (0 to180) and slightly positive during the upwind turn (180 to 360).

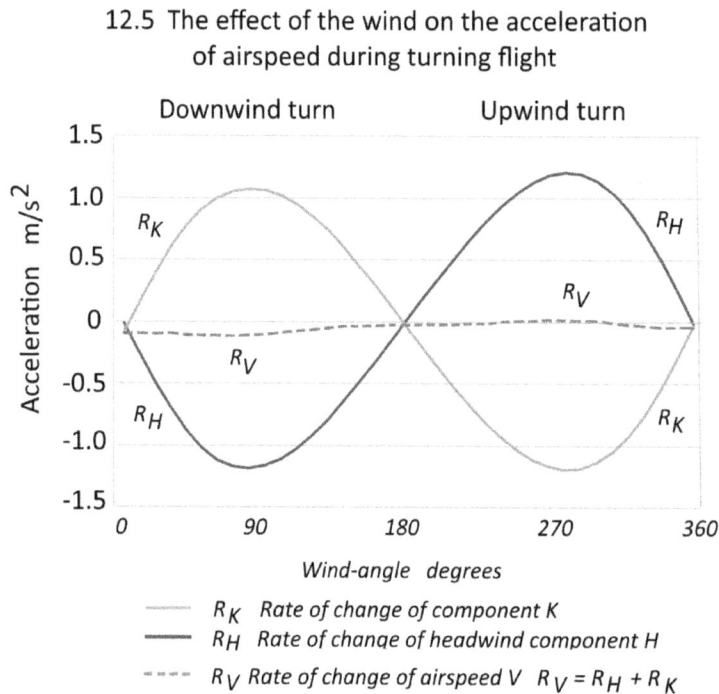

12.5 The effect of the wind on the acceleration of airspeed during turning flight

Note that R_K is zero at the 0, 180 and 360 degree positions because the drift angle is zero and therefore the F_{GT} component is zero. Meanwhile the magnitude of R_H is non-zero at the 0,180 and 360 positions because the rate of turn is non-zero at these positions. In other words, R_H and R_K are zero at slightly different values of wind-angle and always have opposite signs. This is not easy to see but the give-away is the non-zero values of R_V at these points.

These calculations can now give a graphical output of groundspeed and airspeed versus wind-angle. See figure 12.6. In this graph, the groundspeed varies by twice the wind-speed from **V-W** going upwind, to **V+W** going downwind. The airspeed shows a slight variation with the speed reducing during the downwind (leeward) turn and increasing during the upwind (windward) turn. See figure 12.7 with an expanded **y** axis, showing the overall loss of airspeed. This is not the loss of airspeed due to the increased load factor. It is purely the effect of the centripetal force. These results show that the effect of turning a full 360 degree circle in a wind is that the acceleration of airspeed is non-zero except at only two points and therefore the airspeed is <u>not</u> exactly constant.

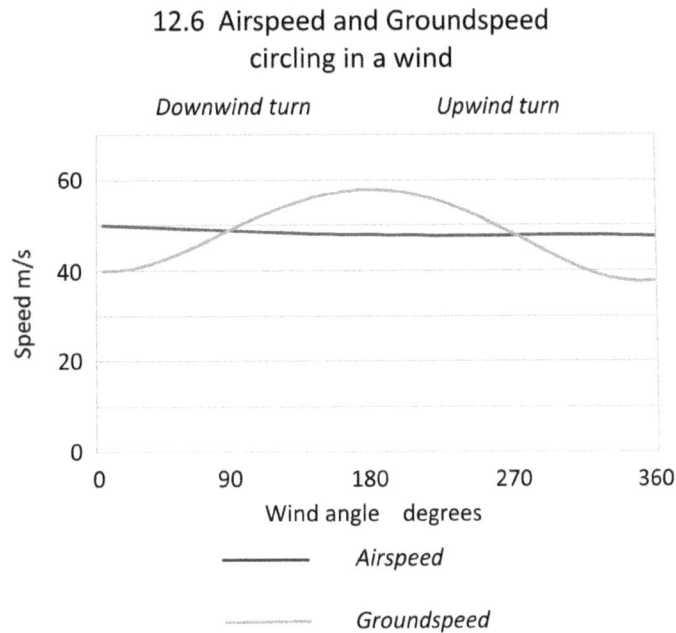

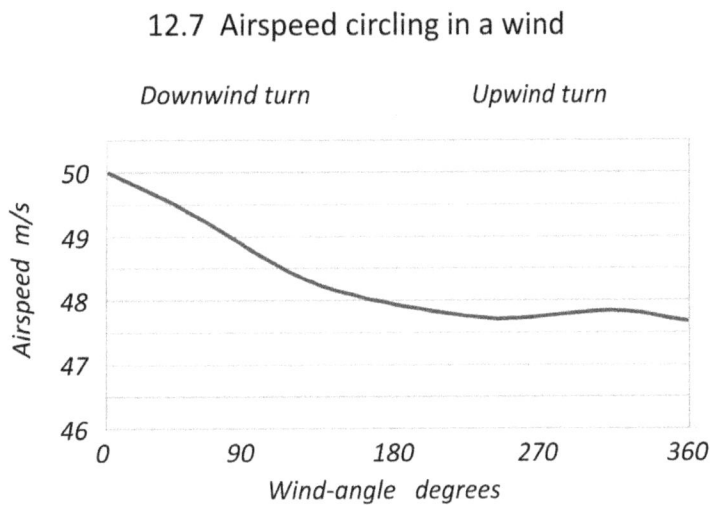

12.6 Airspeed and Groundspeed circling in a wind

Downwind turn　　　　*Upwind turn*

------- Airspeed

········· Groundspeed

12.7 Airspeed circling in a wind

Downwind turn　　　　*Upwind turn*

DYNAMIC SOARING DISSECTED

The overall loss of energy

Does this mean that average airspeed is the same as circling in still air? No. When airspeed varies cyclically, the drag also varies. Drag is a square law, therefore the drag increase associated with a given increase in airspeed is greater than the drag reduction from the same reduction of airspeed. This can be seen in figure 12.7 where the starting airspeed is 50 but the average airspeed is about 49.4 and there is a net loss of airspeed overall. So, if airspeed varies cyclically, even by a small amount, when circling in a wind, then average drag will be greater and average airspeed will be less compared with circling in still air. This means that there is a loss of energy, which could in the end, be speed or height or both.

In practice, airspeed *could* be made to be exactly constant by varying thrust or accepting a variation of height. For a genuinely constant airspeed, there must be a loss of height or, for a glider a slightly increased rate of descent. The effect is the same as when flying in turbulence. For a given thrust setting, the average drag load in turbulence is slightly greater than in still air and the average airspeed is slightly less.

My personal experience of this was during 13 years as a flying instructor on Piper Tomahawks. When circling waiting to land, I used to wonder why, on some days, the aircraft seemed to be labouring and needed an extra 50 rpm to maintain height and airspeed. There are, of course, many possible reasons for this, air-density, aircraft weight and so forth and it is difficult to reproduce the effect by comparing flights on calm and windy days. But looking back, I think the effect of the wind, as described here, was the most likely cause.

Summary

During centripetal acceleration with nominally constant thrust or lift/drag ratio, the effect of the wind is to cause an angle of drift. This causes different tangential and centripetal acceleration components relative to the air-velocity and the ground-velocity.

The air-centripetal force $\mathbf{F_C}$ produces ground-tangential component $\mathbf{F_{GT}}$ and ground-centripetal component $\mathbf{F_{GC}}$ due to drift-angle $\mathbf{d}$. These produce tangential and centripetal accelerations of ground-velocity $\mathbf{G}$ and slightly unequal and opposite accelerations of components $\mathbf{H}$ and $\mathbf{K}$.

Force component $\mathbf{F_{GT}}$ causes ground-tangential acceleration, seen as a decrease of groundspeed in the windward turn and an increase of groundspeed during the leeward turn.

However, because $\mathbf{R_H}$ and $\mathbf{R_K}$ have opposite signs and $\mathbf{R_H}$ is caused by a constant rate of turn, while $\mathbf{R_K}$, as a tangential acceleration, is affected by inertia, there will always be a slight tendency, however small, for airspeed to increase in the windward turn and decrease in the leeward turn.

During circling flight in a wind, if height is exactly constant there will be a very slight variation of airspeed with an increase in total drag losses which will be seen as a slight reduction of average airspeed compared with still-air conditions.

This will only be significant when the wind-speed is a large proportion of the airspeed; which may well have been the case in the early days of aviation when aircraft were slow and under-powered. This in turn, may have led to the warnings about the dangers of the downwind turn.

The gain of airspeed in the windward turn is a <u>very small</u> effect. Nevertheless, it leaves the possibility that airspeed might increase in the windward turn even with an unbalanced drag force. For the albatrosses, the smallest gain of energy during each and every dynamic soaring manoeuvre means they can keep flying for thousands of kilometres.

Now, having descended into this particular rabbit-hole, are you are feeling more like a puffin than a petrel? Well, things get really complicated in the next chapter, as we consider how the wind affects a glider which has both centripetal and drag forces.

Chapter 13

The effect of the wind on a glider

The effect of the wind on a glider

In the previous two chapters, the effect of the wind during aircraft acceleration and its effect on an aircraft in a turn was explained. In a simple 360 degree turn the overall effect is a loss of energy in the form of either speed or height. How then does an albatross turn this effect to its advantage and maintain height and speed?

In chapters 8 and 9 we saw what albatrosses typically do. They do not fly full 360 degree circles; they fly alternately left and right arcs of a circle, close to the crosswind direction where the rate of change of groundspeed and headwind components are at a maximum; there lies a clue as to how they turn the effect of the wind to their benefit.

A glider at constant airspeed is normally propelled by gravity, with its drag losses equivalent to a loss of potential energy in the form of a loss of height at a constant rate of descent. Consider now what happens to a glider which is maintaining height, losing speed and turning so that there are both centripetal and tangential accelerations.

Forces in the Windward Turn

See figure 13.1 which shows two views of the same situation. The wind is from the left and the bird/glider is turning left in a windward turn, maintaining height and therefore losing speed. Again, only the horizontal components are shown.

As well as centripetal force F_C, there is additionally an unbalanced drag force F_D and the vector sum of these forces is the horizontal resultant force F_R. F_R depends on the angle of bank which determines centripetal force F_C and on the lift/drag ratio which determines the drag force F_D.

Whereas in chapter 12 we used only centripetal force F_c, now we use horizontal resultant F_R to provide the ground centripetal F_{GC} and ground tangential F_{GT} force components relative to the ground-velocity.

$$\overrightarrow{F_D + F_C} = F_R = \overrightarrow{F_{GT} + F_{GC}}$$

13.1 Force components
relative to air and ground-velocities
for a glider in a windward turn

Two plan views of the same position

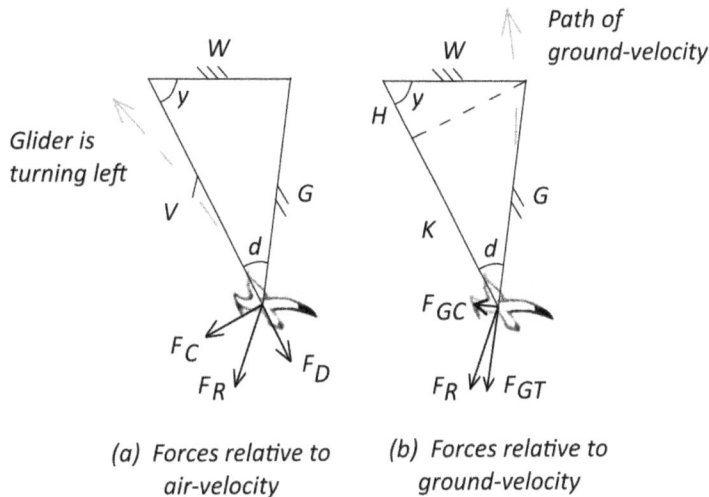

(a) Forces relative to
air-velocity

(b) Forces relative to
ground-velocity

$V = H + K$

F_D is an unbalanced drag force
Centripetal force F_C depends on angle of bank

$$\overrightarrow{F_D + F_C} = F_R = \overrightarrow{F_{GT} + F_{GC}}$$

Force components F_{GT} and F_{GC} produce tangential and centripetal accelerations of the ground-velocity G. F_{GC} provides the rate of turn of the ground-velocity but that just exists in the background and is not used to calculate R_V.

Now, here is the most difficult thing to understand and quite counter-intuitive. It would appear that drag force F_D must make airspeed V reduce but this is not so. Think back to chapter 8; in the albatross windward turn we saw that the groundspeed reduced but the airspeed did not reduce. This tells us that the drag force does not act directly on the airspeed but indirectly through the ground-velocity. This is because airspeed V is not the actual speed of the bird; groundspeed G is the actual speed and the rate of change of G is caused by component $\mathbf{F_{GT}}$. A component of the tangential acceleration of groundspeed $\mathbf{G}$, modified by the angle of drift, then appears as $\mathbf{R_K}$ the acceleration of component $\mathbf{K}$. To repeat once again, the total force in the form of horizontal resultant $\mathbf{F_R}$ is the same in both air and ground frames of reference but with different tangential components and centripetal components in each frame. In the air frame of reference, the centripetal component $\mathbf{F_C}$ provides the rate of turn which determines $\mathbf{R_H}$ and which is unaffected by $\mathbf{F_D}$ at any particular wind-angle.

It is complicated, is it not? The bird/glider is clearly losing speed, that is to say groundspeed, because it is turning upwind, but is it losing airspeed? Component $\mathbf{K}$ is reducing in this windward turn but, as we have seen before, component $\mathbf{H}$ is increasing because wind-angle $\mathbf{y}$ is reducing. In the next section we will see that in certain circumstances, even for a glider, $+\mathbf{R_H}$ is greater than $-\mathbf{R_K}$ and airspeed will increase.

Optimising angle of bank for a gain of airspeed in a windward turn

In dynamic soaring the albatrosses fly a characteristic sinuous, undulating gliding flight path with alternating windward and leeward turns. The windward turn has a shallow angle of bank and constant height above the surface, in ground effect. There is an unbalanced horizontal resultant force which causes the groundspeed and component $\mathbf{K}$ to reduce, whilst a rate of turn causes the headwind component $\mathbf{H}$ to increase. Is there a special case where the airspeed is constant? This may sound improbable but the answer is yes and this is exactly what albatrosses do during dynamic soaring. There is no absolute proof of this; it depends on how the manoeuvre is flown, particularly on the angle of bank, so it is necessary to work an example.

From GPS tracking, we know that during dynamic soaring, albatrosses typically fly on crosswind headings plus and minus 20 to 30 degrees. For a snapshot of one point in the windward turn, say turning through a crosswind heading at a wind-angle of 90 degrees, we can calculate values of $\mathbf{R_H}$, $\mathbf{R_K}$ and $\mathbf{R_V}$ using a range of angles of bank.

We can use some data typical of an albatross: air-velocity 20m/s, wind-velocity 15m/s, lift/drag ratio 20. The principle variable is the angle of bank; we will use a range of 0 to 12 degrees. See Appendix 1 for the calculations.

Remember that the windward turn is a turn through a range of wind-angles passing through the crosswind wind-angle. Figure 13.2 represents one point in the windward turn, when the wind-angle is about 90 degrees and the angle of drift is at a maximum. The horizontal axis is angle of bank and the vertical axis is tangential acceleration. The angle of bank is negative because, in a windward turn, the angle of bank is on the opposite side to the angle of drift. The $\mathbf{R_H}$ line gives the rate of increase of the headwind component $\mathbf{H}$ due to the rate of turn caused by centripetal force $\mathbf{F_C}$. As the angle of bank increases, the rate of turn increases and the rate of increase of the headwind component increases. The $\mathbf{R_K}$ line gives the rate of change $-\mathbf{R_K}$ of component $\mathbf{K}$ due to the effect of the force component $\mathbf{F_{GT}}$ and the angle of drift. The effect of $\mathbf{F_{GT}}$ is retarding and the greater the angle of bank the greater is force $\mathbf{F_R}$ and consequently the retarding effect of $\mathbf{F_{GT}}$. The $\mathbf{R_V}$ line gives the total acceleration of the airspeed, the sum of the other two lines.

$$\mathbf{R_V} = \mathbf{R_H} + (-\mathbf{R_K})$$

13.2 Glider windward turn
Acceleration vs angle of bank

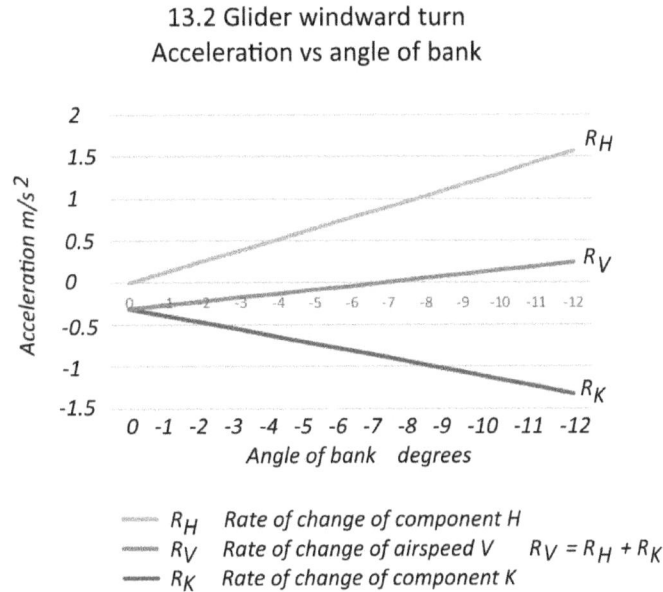

The acceleration R_V of airspeed is found to increase with angle of bank. It increases from a negative value (reduction of airspeed) at zero angle of bank; through zero acceleration (constant airspeed) at an angle of bank of about 7 degrees, to a positive value (increasing airspeed) at greater angles of bank.

This means that if an albatross maintains an angle of bank of about 7 degrees, turning to windward through a wind-angle of about 90 degrees, then it will achieve constant airspeed. This is because, with a lift/drag ratio of 20, the unbalanced drag force is relatively small, while the rate of turn is relatively quick. It is possible to repeat this calculation for all wind-angles between about 130 to 050 (crosswind +/- 40) and show that the albatross can maintain airspeed and height in a windward turn for about 15 seconds.

This explains what can be seen in film of albatross dynamic soaring when they are skimming the surface in a shallow-banked turn without gaining height and therefore without any wind-gradient effect. They appear to be maintaining airspeed because if they did not then they would have to pitch up to increase their angle of attack to maintain lift. Without the dynamic soaring effect, the albatross, at an airspeed of 20 m/s with a lift/drag ratio of about 20 and therefore a rate of descent of 1m/s, from a height of 2m will hit the surface in only 2 seconds.

Controlling airspeed in the windward turn

Albatrosses fly fast compared to most birds but not compared to manned aircraft. Despite this, they can achieve a sufficient rate of turn at a modest angle of bank in wind-speeds which are a large proportion of their airspeed and where the maximum angle of drift is quite large.

Figure 13.2 shows that once constant airspeed is achieved at the optimum angle of bank, small increases and decreases of angle of bank will cause the airspeed to increase or decrease. This is useful for two reasons: firstly, in the windward turn, to take advantage of ground effect, the bird has to climb and descend in order to follow the rising and falling surface of the sea, which would normally make the airspeed fluctuate and increase the average drag loading. By varying the angle of bank, the bird can minimise the fluctuation of airspeed and minimise drag losses. Secondly, although this explanation is predicated on the idea of constant airspeed, in practice it will be necessary to achieve a small gain of airspeed in the windward turn to allow for airspeed losses in the leeward turn.

The Leeward Turn

For a powered aircraft, when thrust equals drag, there is no unbalanced drag force and therefore, as soon as a powered aircraft banks into a leeward turn, there is a component of the centripetal force making the ground-speed increase. For an aircraft without power, that is a glider <u>in level flight</u>, there is an unbalanced drag force contributing to force F_R which makes component F_{GT} retarding at small angles of bank and propulsive at large angles of bank.

Therefore, in the leeward turn, the circumstances of the manoeuvre make a big difference to the outcome. In particular whether a small or large angle of bank is used.

A glider turning downwind with a <u>small</u> angle of bank

See figure 13.3 which shows two views of the same glider (an aircraft with a spluttering, failing engine) at the same moment. In a leeward turn to the right with the wind from the left, the bank angle and the drift angle are both to the right. The glider is <u>maintaining height</u> (but not for long!) therefore F_D is an unbalanced drag force.

13.3 Force components relative to air and ground-velocities for a glider in a leeward turn with a small angle of bank

Two plan views of the same position

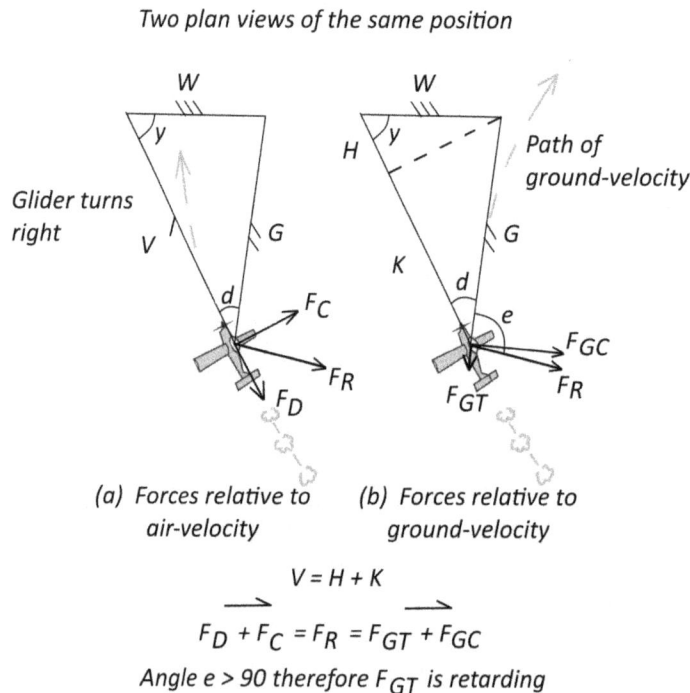

(a) Forces relative to air-velocity

(b) Forces relative to ground-velocity

$$V = H + K$$
$$\overrightarrow{F_D} + F_C = F_R = \overrightarrow{F_{GT}} + F_{GC}$$

Angle e > 90 therefore F_{GT} is retarding

In figure 13.3(a), the force components are oriented to the air-velocity. Drag force F_D combines with centripetal force F_C to create horizontal resultant force F_R. With a small angle of bank, F_C and F_R are relatively small. In figure 13.3(b), the force components are relative to the ground-velocity. Horizontal resultant force F_R resolves into ground-tangential force F_{GT} and ground-centripetal force F_{GC}. The angle e between force F_R and velocity vector G is greater than 90 degrees, therefore ground-tangential force F_{GT} is retarding and causes a loss of groundspeed G and a consequent reduction of component K. (Compare with figure 12.3 which shows the case of a powered aircraft in a leeward turn in which F_{GT} is propulsive to the ground-velocity).

Now see figure 13.4 which shows the same situation but this time two successive positions of the same gliding aircraft in a level leeward turn, turning to the right with a small angle of bank. Component H is decreasing because wind-angle y is increasing. Component K is also reducing because ground-speed G is reducing. Notice the negative rates of change of $-R_H$ and $-R_K$. (Compare with figure 12.3 in the previous chapter). The loss of airspeed is caused both by the aerodynamic forces and because the aircraft is turning downwind. With both H and K reducing, there is a rapid reduction of airspeed V and not a simple incremental change of airspeed.

13.4 Effect of increasing wind-angle for a glider in a leeward turn with a small angle of bank

Plan view of two successive positions

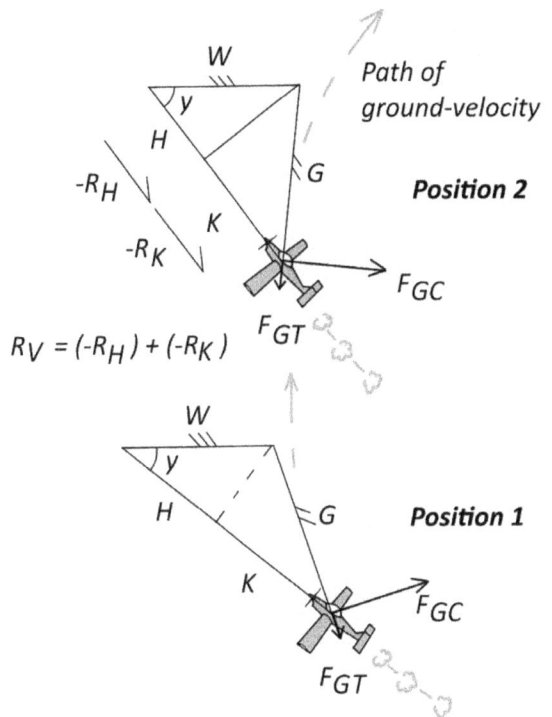

$$R_V = (-R_H) + (-R_K)$$

As wind-angle y increases, component H reduces Force component F_{GT} makes ground-speed G reduce therefore component K reduces

The effect of $-R_H$ is a real loss of airspeed additional to the effect of drag and any mishandling by the pilot.

This condition may occur when a powered aircraft unexpectedly loses power after take-off and the pilot does not get the aircraft nose down sufficiently to maintain airspeed in the glide, whilst trying to turn back to the airfield.

The 'Myth' of the Downwind Turn

From time to time, dating back to the earliest years of aviation, there has been a debate in the aviation community concerning the downwind turn and its role in aircraft accidents. Certainly, stall-spin accidents during a downwind turn following an engine failure, are mostly caused by miss-handling of the aircraft by the pilot. This is caused by the confusing visual effects of increasing ground-speed and drift, creating the impression of increasing airspeed and side-slip, inducing the pilot to apply up-elevator and rudder, leading to a stall and spin. The loss of airspeed in the downwind turn is occurring at the very moment when drag and stalling speed are increasing due to the increasing load factor.

We can now see that there is an actual physical effect on the aircraft, caused by the downwind turn, which will cause the airspeed to decay but the effect is very small. The slight additional reduction of airspeed during the downwind turn may become significant if the aircraft is close to the stalling speed and thus precipitate a stall-spin accident.

You can see this effect when watching a free-flight model aircraft which is trimmed to turn in a circle. As it turns downwind it tends to sink and lose airspeed causing its nose to drop. As it turns upwind it gains airspeed and pitches up again.

When turning downwind, if airspeed is to be maintained, it is necessary for groundspeed to increase. Without power and with a small angle of bank, the only way to increase groundspeed is to use gravity, to lower the aircraft nose and increase the rate of descent. That may sound like stating the obvious but the situation changes when the bank angle increases, as shown in the next section.

A glider turning downwind with a <u>large</u> angle of bank: Modelling the albatross leeward turn.

With a large angle of bank, air-centripetal force $\mathbf{F_C}$ is much greater in magnitude and therefore, horizontal resultant force $\mathbf{F_R}$ is also larger and rotates counter-clockwise, reducing the angle **e**. When $\mathbf{F_C}$ has increased, due to increased angle of bank, but $\mathbf{F_D}$ has <u>not</u> increased, to the point that angle **e** reduces to 90 degrees, then retarding force $\mathbf{F_{GT}}$ reduces to zero and $\mathbf{F_R}$ and $\mathbf{F_{GC}}$ are coincident (not shown). This will only happen if the drag force $\mathbf{F_D}$ does <u>not</u> increase, which can only happen in a low-G wing-over. With a further increase of angle of bank, $\mathbf{F_C}$ and $\mathbf{F_R}$ increase, angle **e** reduces to less than 90 degrees, $\mathbf{F_{GT}}$ becomes a propulsive force and groundspeed **G** increases. (See figure 13.5 compared to figure 13.3).

13.5 Force components for a glider in a leeward turn with a large angle of bank

Two plan views of the same position

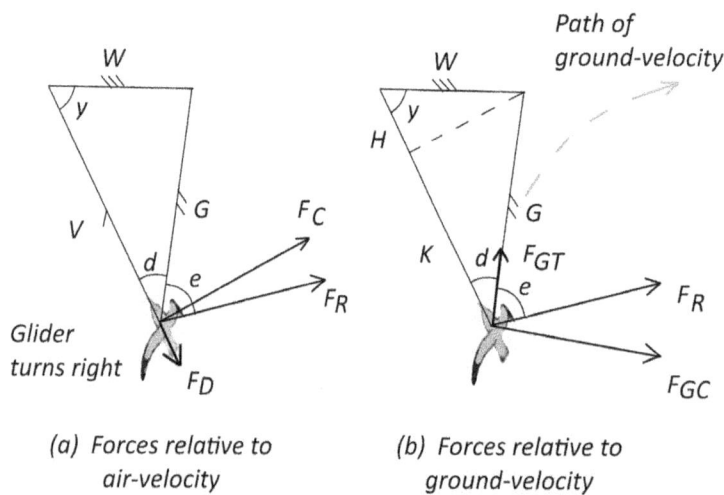

(a) Forces relative to air-velocity

(b) Forces relative to ground-velocity

$$V = H + K$$

$$F_D + \overrightarrow{F_C} = F_R = \overrightarrow{F_{GT}} + F_{GC}$$

F_{GT} is propulsive when angle e is less than 90 degrees

This in turn causes component **K** to increase (**+R$_K$**). (See figure 13.6 showing two successive positions in the leeward turn). Component **H** is still reducing (**-R$_H$**) due to increasing wind-angle **y** but now the outcome will depend on the balance of **-R$_H$** and **+R$_K$**. If the rates of change of **H** and **K** are to balance, this is only possible if **F$_C$** is relatively large and the unbalanced drag force **F$_D$** is small and does not increase by much. This can be achieved by flying the turn as a wing-over at 1G without a big increase in load-factor.

(And by the way, I am not advocating a wing-over as a turn-back manoeuvre following an engine failure. I am just pointing out the difference it makes to an albatross. As they say on TV, 'Don't try this at home!')

The leeward turn as a wing-over

To achieve a large angle of bank without increasing the G-loading, a wing-over is used. A wing-over is an arched turn in which the load-factor is less than that required for a level turn. Instead of a level turn, the aircraft climbs in a partly ballistic path during the first part of the turn and descends in the second part of the turn, trading airspeed for height.

Figure 10.5b showed the forces during a level 45 degree banked turn at a load-factor of 1.4G with the vertical component equal to the weight. A level 60-degree angle-of-bank would require 2G and 70 degrees would need 3G with corresponding increases in the drag force. Figure 10.5c is a 70-degree wing-over at a load-factor of one (1G). Lift equals weight and therefore, there is no increase of drag but the vertical component of lift **L$_V$** is less than the weight.

13.6 Effect of increasing wind-angle for a glider in a leeward turn with a large angle of bank

Plan views of two successive positions

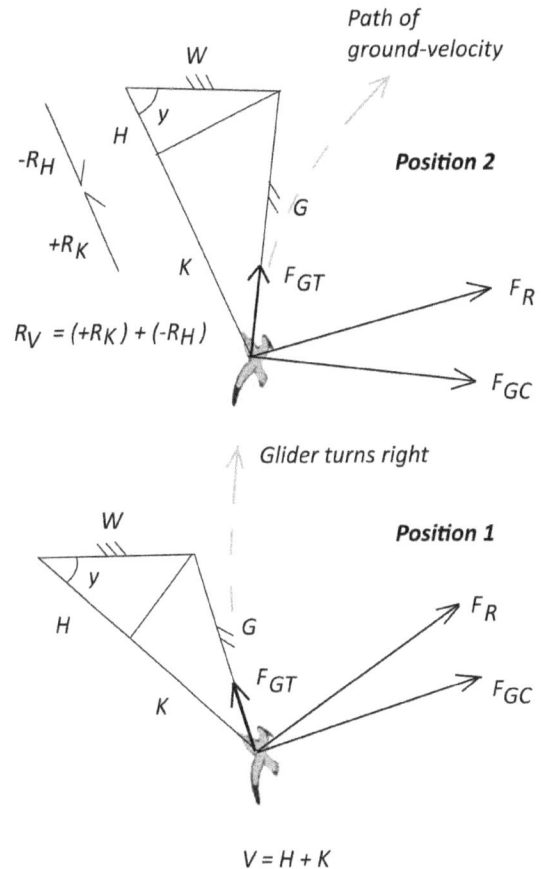

$$R_V = (+R_K) + (-R_H)$$

Glider turns right

$$V = H + K$$

As the wind-angle increases, component H decreases
Ground tangential force F_{GT} makes groundspeed G
increase therefore component K increases

Optimising the leeward turn

Is there once again a condition where the acceleration of component **K** is approximately equal and opposite to the acceleration of component **H** and airspeed is nearly constant? Well, not exactly. In the leeward turn the effect of **-R$_H$** is slightly greater than the effect of **+R$_K$** and there will always be a slight loss of airspeed. Once again we can work an example and this is seen in Figure 13.7, using the same program as produced figure 13.2. Figure 13.7 represents one point in the leeward turn at about 90 degrees wind-angle and using the same data as before but with a wider range of angles-of-bank and with load-factor in the wing-over at a nominal 1G instead of 1/cosine angle-of-bank, which would otherwise be for a level turn. The angles-of-bank are positive because in a leeward turn the angle-of-bank is on the same side as the angle-of-drift.

The **R$_K$** line represents the rate of change of component **K** versus angle of bank. The **R$_H$** line is the rate of change of the headwind component **H** which also depends on the angle of bank but via a different mechanism. The R_V line, the acceleration of airspeed, is the sum of the other two lines, which is seen to be negative at all angles of bank because the effect of $-R_H$ is slightly greater than the effect of $+R_K$. This will result in a loss of airspeed during the leeward turn but the loss does not increase rapidly above about 60-degrees angle of bank.

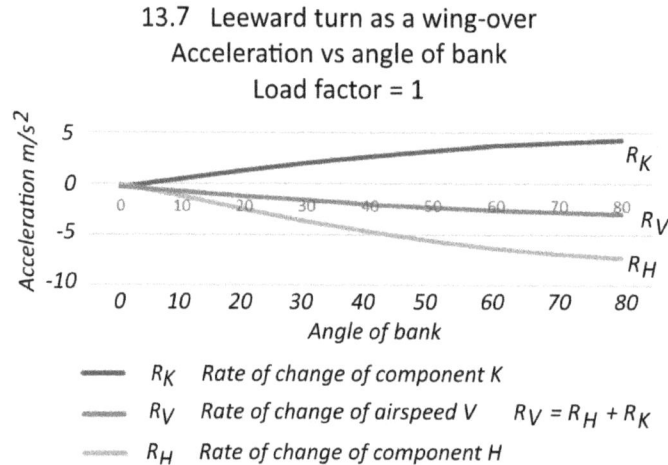

13.7 Leeward turn as a wing-over
Acceleration vs angle of bank
Load factor = 1

R_K R_K Rate of change of component K
R_V R_V Rate of change of airspeed V $R_V = R_H + R_K$
R_H R_H Rate of change of component H

The point here is that, although airspeed is not constant in the leeward turn, a large angle-of-bank can be achieved in a wing-over, enabling a rapid rate of turn without a large load-factor and without a large drag penalty. Also, small changes of angle of bank between 60 and 80 degrees do not make a big difference to the outcome; which is to say there is no particular penalty for an angle of bank of 70 +/- 10 degrees. Control of airspeed in the leeward turn is not possible as there is a small loss of airspeed at all angles of bank, as well as an exchange of airspeed with height during the wing-over. However, this is not critical as the turn is made at height, independent of the motion of the sea surface. This loss of airspeed will, for the albatross, be balanced by airspeed gained in the windward turn. ***This explains why albatrosses habitually fly their leeward turns as 60 to 80-degree wing-overs at about 1G.***

Effect of Load Factor in a level turn

So far as I can tell, albatrosses have never been instrumented to determine load-factor (or airspeed come to that) so it is not known what load-factor the albatrosses use in their wing-overs. It appears that many researchers assume that a steep angle of bank means a large load-factor and they erroneously use that in their analyses of wind-gradient dynamic soaring.

By way of comparison, figure 13.8 shows the acceleration of speed of a glider at different angles-of-bank from 0 to 80 degrees at one point in a level leeward turn with a load-factor of 1/cos angle of bank. Up to about 60 degrees angle of bank, both R_K and R_H increase gradually in opposite senses. Above 60 degrees they increase rapidly with corresponding increases of drag. The sum of R_K and R_H gives R_V the acceleration of airspeed, the middle line, which is negative at all angles of bank and increases greatly at angles of bank greater than about 60 degrees. Compare with figure 13.7, note the different vertical scales. Compared to -2.5 m/s^2 at 1G in a 70 degrees angle of bank wing-over, in a 70 degrees level turn the deceleration is about -8.2 m/s^2 at 2.9G. Therefore, it seems unlikely that the albatrosses are doing this.

13.8 Leeward turn as a level turn
Acceleration vs angle of bank
Lf = 1/cos AOB

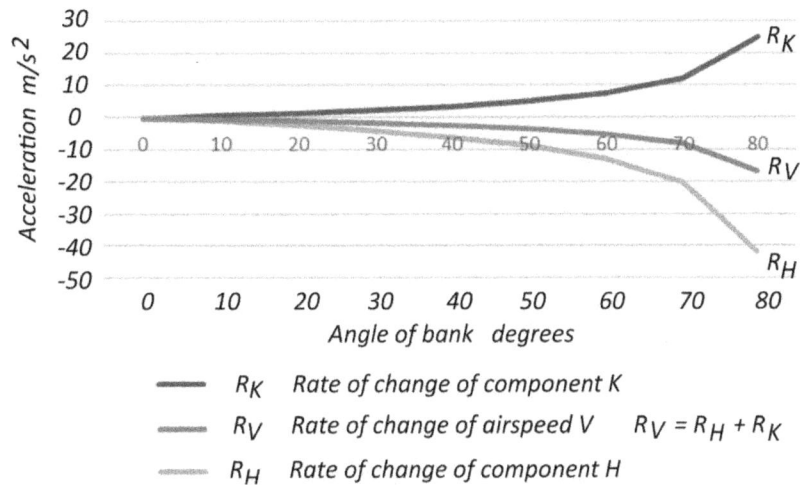

	R_K	Rate of change of component K
	R_V	Rate of change of airspeed V $R_V = R_H + R_K$
	R_H	Rate of change of component H

Theory compared with nature

These results correspond quite well with angles of bank seen in film of albatrosses dynamic soaring. The birds fly windward turns at small angles of bank, close to the surface, taking advantage of ground-effect. Leeward turns are flown as steeply banked wing-overs in which low-G manoeuvring would require minimum effort. Which is just as well, because if dynamic soaring did not occur in nature, any theory proposing it might appear to be preposterous!

It should be noted however that the results are easily modified by making changes to the assumed parameters. For example, in figure 13.7 the load factor in the wingover is assumed to be less than that required for a level turn. There is no guarantee that constant airspeed could be achieved in either the upwind or downwind turns. It depends entirely on how the manoeuvres are flown, especially regarding the angle of bank and the load-factor. Therefore, this is again not a proof but rather an illustration of what is possible.

Modelling albatross dynamic soaring

Doing these calculations for one point in the turn does not prove that dynamic soaring works. To do that requires the whole of the windward and leeward turns to be plotted. However, if the model involves a full 360 degree circle, flown in a wind with a steady angle of bank, the overall effect is a loss of energy in the form of speed or height, as shown in figure 12.7 in chapter 12. That also proves nothing in relation to albatross dynamic soaring. On the other hand, if the range of wind-angles, bank-angles and other parameters are limited to replicate what an albatross actually does, then airspeeds and heights can be compared at the beginning and end of the manoeuvre to determine whether there are gains or losses. A gain of speed or height will demonstrate that the dynamic soaring mechanism is valid. Successful dynamic soaring depends on a minimum wind velocity as well as a particular flight technique so, proving a theory of dynamic soaring is probably impossible. The best that can be achieved is an illustration of animal behaviour; of what a bird does given a certain wind.

So the Windward Turn Theory evolved through an iterative process - seeing what worked. There are a large number of terms and variables and so I have not attempted to reduce the maths to a single equation. It was a bit like juggling with jellyfish. The albatross model is known to work, so its data were used as a starting point. The data used is that of an albatross in typical wind conditions as published in **Avian Flight by J J Videler 2010**. Many of the nominated parameters such as angles of bank, are estimated by observation of film of albatross in flight.

The object of this exercise is to see whether albatross flight can be modelled using the forces acting on the bird, the triangle of velocities and the equations of motion. I have not 'defined a co-ordinate system and solved the equations of motion' as most scientists put it. My co-ordinate system is a hybrid one, centred on the glider but with vertical and horizontal axes defined by the direction of gravity and the relative direction of the wind. An Excel spreadsheet is used with a single set of equations to model both the windward and leeward turns. The end result depends not only on the equations but also on certain parameters such as the range of bank-angles, wind-angles and the wind speed. By accepting a limited range of wind-angles, typically crosswind plus and minus 20 to 30 degrees, we can calculate for each wind-angle, the effect of angle of bank and angle of pitch. The angle of bank will give a centripetal force which will give a rate of turn and therefore a rate of change of headwind or tailwind component. The aerodynamic forces will also control the rate of tangential acceleration, which acts opposite to the rate of change of the head/tailwind component to control the airspeed. The motion of the bird is defined by airspeed, lift/drag ratio, load factor and angles of pitch and roll combined with the wind-speed and wind-angle. The roll and pitch profile is intended to give a nearly flat windward turn and a leeward wing-over. In the spreadsheet, different angles of bank can be tested to see which produce an airspeed, ground-speed and height profile similar to those seen in nature. The important point being to test whether the result gives at least as much airspeed and height at the end of the dynamic soaring manoeuvre as at the beginning.

Results

The three diagrams in figure 13.9 show an example of the results of the calculations. The horizontal axis is crosswind distance in tens of metres, an overall distance of 450m and lasting about 20 seconds. The left side is the windward turn and the right side is the leeward turn, with the bird flying from left to right, reversing the direction of turn at about 370m. The model bird has a mass of 10kg, and a lift/drag ratio of 20. The program works with a starting point of airspeed 20m/sec and height 2m. The wind-velocity is a uniform 10m/s with no wind gradient or any vertical motion of the wind. A profile of angle of bank (10 degrees left in the windward turn and 70 degrees right in the leeward turn) and angle of pitch between 0.5 degrees down and 8 degrees up is then applied at 10m intervals in the crosswind direction giving time intervals of about 0.5 seconds. The program uses the triangle of velocities, the laws of motion and the forces acting on the bird, to calculate the load factor, the centripetal and tangential accelerations, the rate of turn, airspeed and wind-angle (heading relative to the wind), groundspeed and track, and the rate of climb or descent and height.

Heading and Track

Figure 13.9a is a plan view showing the heading of the bird and its path through the air together with the track over the ground. The heading is the same as the wind-angle with the wind coming from the top. It shows the range of wind-angles, turning left during the windward turn from 120 through 90 to 60 degrees and then turning right during the leeward turn from 60 through 90 to 120 degrees.

13.9a Dynamic soaring model
Air/ground plot

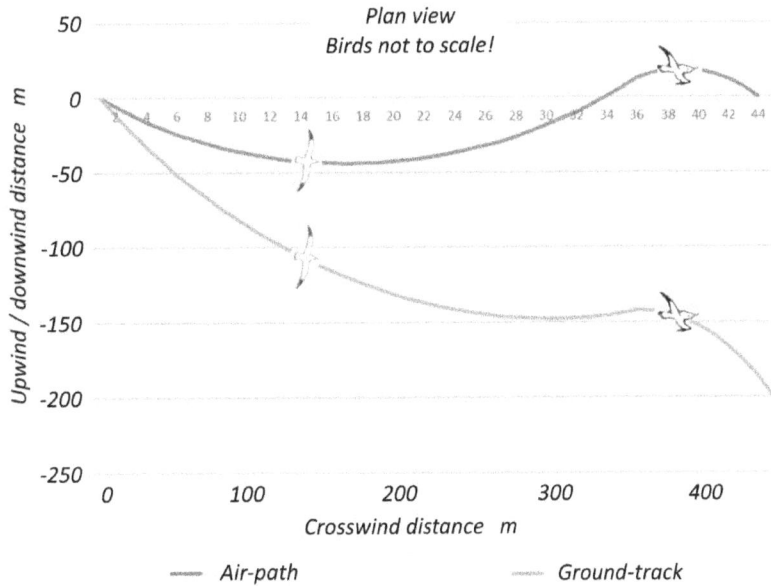

Plan view
Birds not to scale!

Upwind / downwind distance m

Crosswind distance m

— Air-path   ---- Ground-track

The gap between the ends of the lines on the right is the distance lost downwind, about 200m, the effect of the wind. The same bird is shown at two places on each line which are effectively the same places and times on both lines. The bird is not drawn to scale but does illustrate the angle of drift relative to the ground track. The motion of the bird model is similar to the track shown in the GPS data in chapter 8 but with a longer windward turn at a lesser angle of bank in a presumed lighter wind.

Airspeed and Groundspeed

Figure 13.9b shows variation of airspeed and groundspeed against distance flown. During the windward turn, the left side of the diagram from 0 to 370m, the airspeed starts at 20 m/s and then increases to about 26m/s, while the groundspeed begins at about 27m/s and decreases to about 22m/s. During the leeward turn from 370m to 450m, the airspeed decreases as the bird gains height in the wing-over and then increases slightly to about 27m/s as the bird descends, leaving a net gain of airspeed. The groundspeed in the leeward turn decreases slightly as the bird starts to climb and then increases as it turns and descends with a net gain of groundspeed to 35m/s at the end.

13.9b Dynamic soaring model
Speed

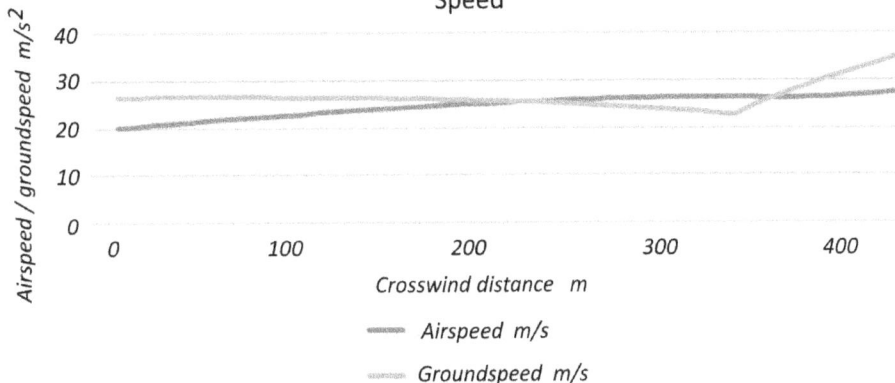

Airspeed / groundspeed m/s²

Crosswind distance m

— Airspeed m/s
---- Groundspeed m/s

Height

Figure 13.9c shows height in metres against distance in tens of metres, with the height scale expanded for clarity. The height at the beginning is 2 metres. The pitch profile in the windward turn gives a slight loss of height and then a slightly greater gain to about 6m. The bird must gain height before reversing the bank angle because of its 2 to 3m wing-span. The leeward turn is then a wing-over rising to about 8 m before descending to 5m, leaving <u>a small gain of height overall</u>. Notice that from about 360m to 380m the model albatross gains both height and groundspeed just as seen in the data from actual albatross GPS tracking in chapter 8. The small kink in the line at 360m is the point at which the model albatross reverses direction between the windward and leeward turn. Bear in mind the relatively coarse interval of 10m between data points.

13.9c   Dynamic soaring model
Height

Side view

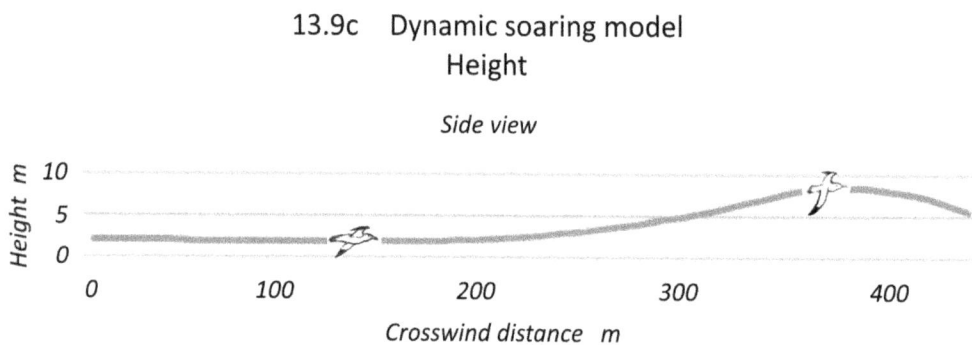

These results are an illustration of how it is possible for an albatross to dynamic soar in a uniform horizontal wind whilst maintaining average airspeed and height. It requires the typical profile of roll and pitch angles which albatrosses are seen to perform. It works for a variety of similar airspeed and wind-speed combinations but any other profile results in a loss of airspeed or height and it breaks down completely if radically different combinations are used, for example it does not work for flying directly downwind or in a 180 degree turn. Note for comparison, that in straight gliding flight with a lift/drag ratio of 20, at constant speed, the model glider will descend from 2m to the surface in a distance of only 40m whereas, in the windward and leeward turns shown here, the bird is able to fly more than 400m without losing overall airspeed or height.

Summary

The effect of the wind on an aircraft in flight is to generate an angle of drift. During accelerated flight, the tangential and centripetal components of force and acceleration have directions relative to the air-velocity and the ground-velocity, which differ by the angle of drift. During accelerated flight, the airspeed can be thought of as comprising a groundspeed component and a head or tail-wind component.

The effect of the aircraft rate of turn, relative to the wind direction, is to create a rate of change of the head or tail-wind component and a consequent acceleration of the airspeed. During a turn, the acceleration of airspeed is the sum of the acceleration of the headwind component and the acceleration of the groundspeed component.

During a windward turn, the aerodynamic force component tangential to the ground velocity causes the ground-speed to reduce. At the same time, the headwind component is increasing (or the tailwind component is decreasing) due to the rate of turn, which will tend to make the airspeed increase.

In the leeward turn the tangential force component makes the groundspeed increase while the headwind component reduces (or the tailwind component increases) tending to make the airspeed reduce.

In normal level turns without unbalanced drag forces the slight variation of airspeed in the windward and leeward turns causes a small drag loss compared with turning in still air.

If an un-powered aircraft maintains height in a leeward turn at a small angle of bank, the loss of airspeed is increased by the reducing headwind component.

The tendency to lose airspeed in the leeward turn contributes to the circumstances which lead to stall-spin accidents unless gravity is used to assist acceleration with consequent loss of height.

Albatrosses are able to maintain or even increase airspeed during windward turns when the rate of increase of the headwind component exceeds the rate of decrease of the ground-speed component.

Albatrosses are able to minimise their airspeed losses during a leeward turn by flying a wing-over at a large angle of bank but with a small load-factor. This, combined with a large angle of drift, means that the aerodynamic force component tangential to the ground velocity is propulsive and causes the ground speed to increase. This only works close to a crosswind heading and only if there is little change to the drag force.

The two special cases of nearly constant airspeed in windward and leeward turns form the circumstances in which albatrosses are able to maintain average height and airspeed during dynamic soaring. When the albatross has achieved a rate of turn that gives approximately constant airspeed in the windward turn, that airspeed can be controlled by varying the angle of bank.

This kind of low-G crosswind dynamic soaring does not require a wind gradient. It always has a downwind drift angle and does not work with circles or 180 degree turns. However, in chapter 9 we saw that albatrosses can readily dynamic soar upwind. How is that possible? To explain this, in the next chapter we will look at how RC model gliders achieve high speeds in circling dynamic soaring flight and then come back to the albatrosses in chapter 15.

Chapter 14

RC model gliders dynamic soaring

Introduction

Radio-control model glider pilots have found a new way of flying in the lee of a hill. Confusingly, this is also known as dynamic soaring but it is not exactly the same as what the albatrosses are doing. Normally, soaring is done on the windward side of a hill to take advantage of the up-draughts as the wind blows up and over the hill; the lee side of the hill is usually avoided because of down-draughts and turbulence. However, it has been found that it is possible to maintain height, position and speed in a circular flight path, on the lee-side of a hill, whenever there is a marked shear boundary between the wind blowing over the top of the hill and a relatively still area below. This technique was developed in the 1990's by Joe Wurts in the hills of California and subsequently developed by Spencer Lisenby and others.

In the right conditions, huge airspeeds exceeding 500 mph have been achieved, as recorded by radar speed detectors, or more modest airspeeds have been sustained in lighter winds, downwind of trees or sand dunes. It appears that this flying technique has been invented or discovered by experienced model glider pilots who have extended their skill and knowledge in an intuitive kind of way. The manoeuvre was not copied from nature; no birds are known to behave in exactly this way although some aspects of the manoeuvre are seen in avian displays.

The shape of the manoeuvre is quite different to what albatrosses do, although since the models and the birds are both gliders, they must be bound by the same rules of dynamics. Any description or mathematical modelling of these manoeuvres must use the same underlying principles. If a model can circle over a fixed position in a wind, it is effectively moving upwind. Therefore, the dynamics of circling flight must be essentially the same as what the albatrosses do when dynamic soaring upwind. We will return to the albatrosses in the next chapter.

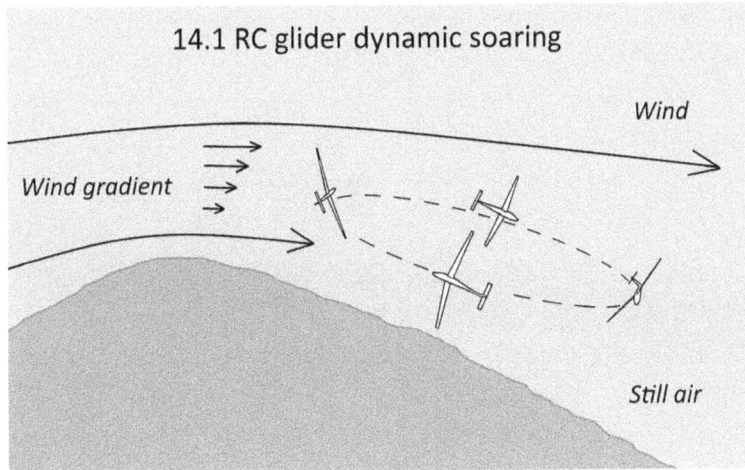

14.1 RC glider dynamic soaring

Wind-gradient theory?

Again, the wind-gradient theory of dynamic soaring is cited as the mechanism in use. It goes something like this: The glider turns downwind and descends through the shear boundary, groundspeed is preserved and airspeed increases. The circuit is completed in still-air below the shear boundary with a little loss of speed. The model turns upwind and climbs through the shear boundary into the fast moving air, airspeed increases again and the cycle is repeated.

So what is wrong with that explanation? In this form of dynamic soaring both airspeed and ground-speed must increase; we know this from the radar speed detectors which are measuring groundspeed. But the theory does not say how groundspeed increases. Also, the wind gradient has maximum effect only if the glider penetrates the shear boundary on upwind or downwind headings. If they are penetrating the shear boundary on approximately cross wind headings or in a turn, the effect of the wind-gradient is much reduced. This form of dynamic soaring is not the same as the low-G dynamic soaring that the albatrosses do and nor is it completely explained by the wind-gradient theory. In order for both airspeed and groundspeed to increase there has to be a force acting in the direction of motion to achieve that acceleration and also to overcome the increasing drag with increasing airspeed. The wind gradient theory does not show such a force. The wind is an obvious source of energy but how is energy transferred from the wind to the glider?

Adapting the Windward Turn Theory to RC dynamic soaring

What happens when an RC glider circles, climbing and descending in a deep wind-gradient; bearing in mind that the pilot cannot see where the wind-gradient or the shear boundary begins and ends? The Windward Turn Theory can be applied to RC dynamic soaring but the same basic mechanism works differently in the two cases because the albatrosses fly a low-G, undulating flight path and the RC gliders fly a high-G circular flight path. Such a mechanism is also needed to explain how albatrosses dynamic soar upwind.

All of the diagrams in this chapter are from the same single set of equations used to plot the illustrations of the albatross crosswind and upwind flight paths; only the starting data and the shape of the manoeuvre are different. The model of albatross dynamic soaring was made with only two angles of bank, one for each of the windward and leeward turns. The RC glider flight manoeuvre is more complicated than those of the albatrosses because of the need to adjust the various parameters to ensure the end point is close to the starting point and does not drift downwind. To achieve this, the angle of bank and therefore the load-factor must be varied continuously around the 360 degrees of the circle simulating the pilot's control inputs.

The program calculates and plots one 360-degree circle at 10 degree intervals of wind-angle, starting at the lowest point, on a crosswind heading in the middle of the windward turn. It uses airspeed, angle of bank and angle of climb, load factor, mass, lift/drag ratio, the wind and drift angle and the aerodynamic forces and gravity. The results, in numerical or graphical form, can be inspected to see how much speed or height is gained. The final result is achieved by an iterative process of adjusting the inputs and seeing what works. If airspeed is gained, that value can be manually entered as the starting airspeed and the next circle plotted.

And this is the point: I am not simply assuming that airspeed gained in the wind-gradient equals airspeed lost due to drag. I am applying the aerodynamic and gravitational forces, together with the rate of change of the headwind component and discovering *whether or not* there is an increase of speed. The results show that the manoeuvre is not energy neutral, as some people think. Rather, it shows that there is a net energy gain in the form of increased airspeed and groundspeed. The energy is derived from the wind using the forces acting on the aircraft which are derived from momentum given to the air as explained in the previous chapters.

The Wind gradient

The wind encountered by the aircraft depends on the aircraft height within the wind-gradient layer, starting at the lowest point of the circle and climbing and descending through the wind gradient. To generate a wind profile in the spreadsheet, the wind used is the product of the maximum wind and the logarithm base 10 of the aircraft height which gives a wind versus height profile as seen in figure 14.2. The imaginary hill-top producing this effect is some arbitrary height and distance upwind.

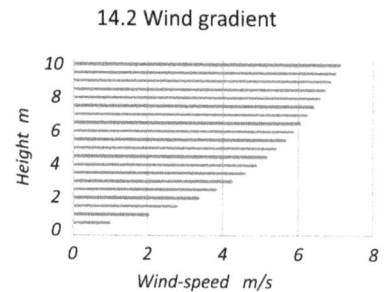

14.2 Wind gradient

The height gained by the aircraft will depend on the angle of climb and the airspeed. However, the angle of climb is assumed to be small, approximately 5 degrees, to avoid introducing extra geometric complications.

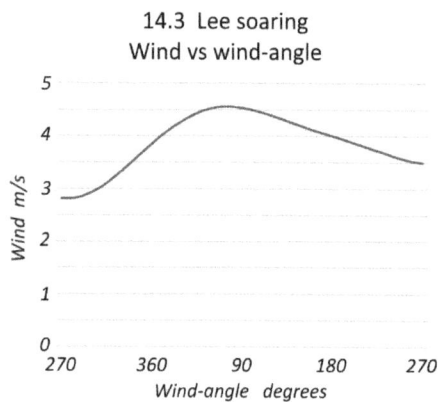

14.3 Lee soaring
Wind vs wind-angle

Wind

Figure 14.3 shows the actual wind experienced by the glider as it climbs and descends through the wind gradient. The final wind is greater because the glider ends up with a slight gain of height.

Height

The height of the aircraft during the circle is shown in figure 14.4. The circle starts at an arbitrary 2m on the crosswind heading of 270 degrees and peaks at about 4m just before the opposite crosswind heading between 60 and 090 degrees. It ends with a slight height excess. Avoiding negative heights makes operation of the spreadsheet easier and is, of course, a necessity to avoid losing energy to gravity. If the maximum wind-speed is

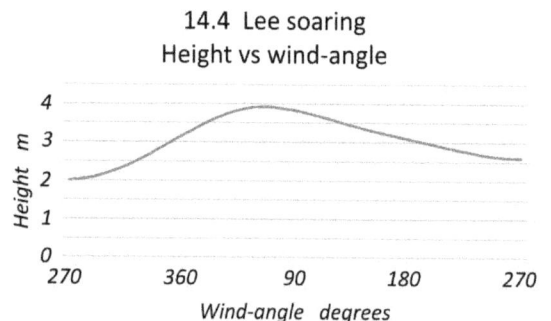

14.4 Lee soaring
Height vs wind-angle

increased and the circle is allowed to drift downwind of its starting position, it will generate negative heights and this is avoided by increasing the bank-angle as the glider turns downwind as seen in the air-plot.

Angle of bank

14.5 Lee soaring
Angle of bank vs wind-angle

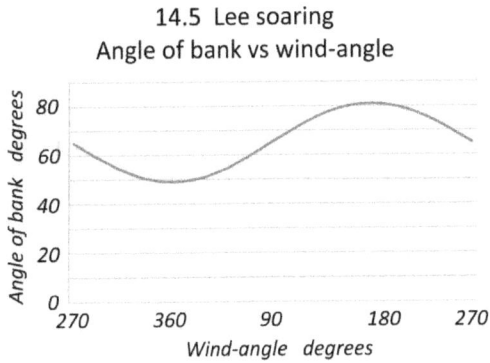

In real model flying, the angle-of-bank is under the control of the pilot and must be varied to keep the circular flight-path of the aircraft in approximately the same location over the ground. The angle of bank is least on upwind headings, between 270 through 360 to 090 wind-angles and greatest on downwind headings.

In the simulation, the aircraft angle-of-bank is generated by a sine function depending on the wind-angle, that is the position of the glider in the circle, and on bank-angle amplitude. The bank-angle amplitude is the difference between the maximum and minimum bank-angles. In this diagram the bank-angle varies from 50 to 80 degrees. A phase-angle is added to the wind-angle to position the maximum angle-of-bank just before the 180 degree wind-angle which is the downwind heading. (Figure 14.5)

Each time a change is made, the ground plot is inspected and the bank angle adjusted, to ensure the aircraft ends up approximately back at its ground starting position to ensure it does not drift downwind.

Load factor

The variation of load-factor in figure 14.6 corresponds to the variation of bank-angle but is factored to be greater than that for purely level flight. A minimum of 1.5G is achieved just before the up-wind position and a peak of 8G just before the lower downwind heading. The idea is to generate an artificially high centripetal force and acceleration of groundspeed. The model assumes constant drag during each circle but in reality the increased load factor must lead to increased drag but this is difficult to allow-for. The faster the model flies the greater the load factor needs to be to keep the circle in position. In real life it is thought the models are achieving extremely high G loads (and frequently break up in flight!)

14.6 Lee soaring
Load factor vs wind-angle

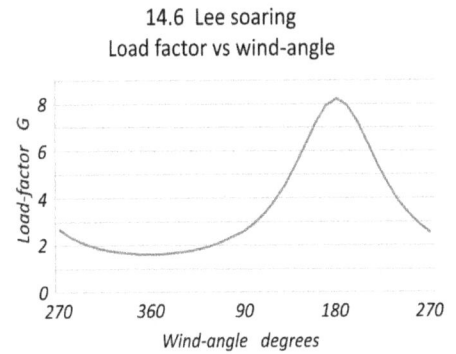

Air and Ground plots

The Air-plot (figure 14.7) is the path of the aircraft through the air, starting at the zero point, on a heading of 270 degrees turning right. It illustrates, during the second half of the circle, the increased rate of turn and reduced turn radius caused by the increased angle of bank. This is stretched out by the wind as seen in the ground plot.

Bear in mind that the intervals between the data points are NOT time intervals but are wind-angle intervals of 10 degrees; so that the gaps between data points do not represent the speed. This means that the centre of the circle is not in a fixed position but is itself moving in a circle (See the slingshot analogy later).

14.7 Lee soaring
RC glider air-plot
Plan view

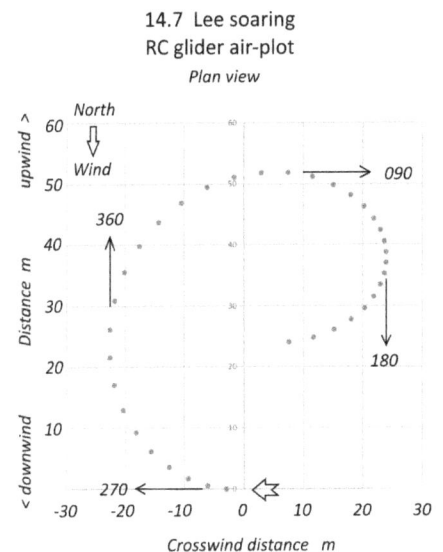

14.8 Lee soaring
RC glider ground plot

Plan view

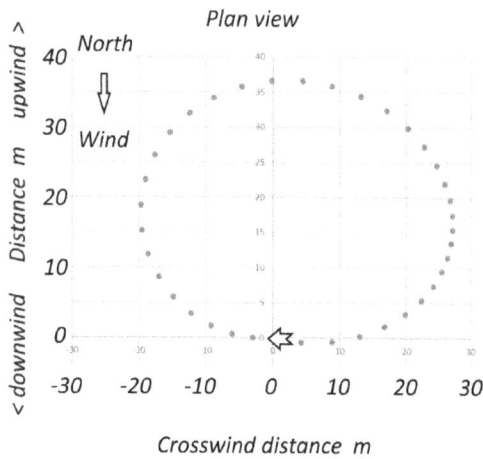

Crosswind distance m

The next diagram (figure 14.8) shows the Ground-plot, starting at the zero coordinates, with a wind-angle (heading) of 270 degrees, turning right. The wind is from the North. The diameter of the circle is 35 to 48m and ends up slightly downwind of the starting point. The bank-angle is least on the upwind heading on the left side of the diagram and greatest on the downwind heading on the right side of the diagram.

Note that in these diagrams the circle starts at 270 degrees turning right (upwind) through a full 360 degrees. The graph is the same if the starting heading is 090 degrees turning left (upwind).

Airspeed and ground-speed

The speed diagram (figure 14.9) shows the airspeed and ground-speed during one circle. This result is achieved by adjusting the angle of bank and load-factor inputs until the desired result is obtained. It clearly shows a gain of speed overall.

The airspeed is a nominal 20 m/s at the start, at the lowest point of the wind-gradient where the wind-speed is zero and therefore the drift-angle is zero and the groundspeed is the same as the airspeed.

The groundspeed reduces in the first quarter, as you would expect turning into an increasing headwind and then increases during the downwind or leeward turn from 360 to 180, the middle half of

14.9 Lee soaring
Airspeed and groundspeed vs wind-angle

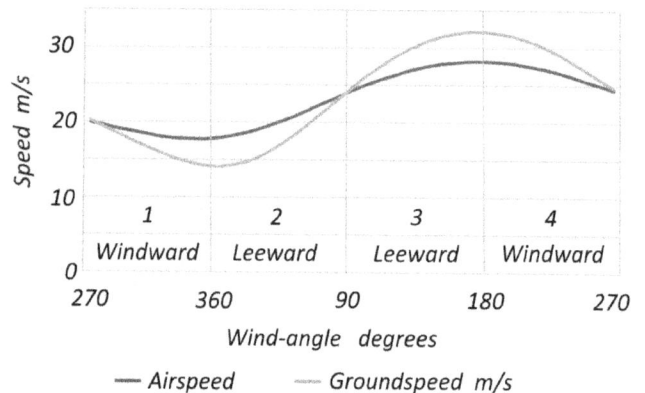

Wind-angle degrees

— Airspeed ⋯ Groundspeed m/s

the diagram, as the aircraft gains and loses height. This is because the drift angle is relatively large and produces a large tangential force making the groundspeed increase. During the fourth quarter from 180 to 270, the ground-speed starts to reduce again but ends up about 4 m/s more than the starting ground-speed.

Airspeed is affected by both the acceleration of the ground speed and by the rate of change of the headwind component but in the relatively high-G RC dynamic soaring, aerodynamic force is dominant and therefore R_K is dominant giving an increase of airspeed in the leeward turn. However, R_K depends on the aerodynamic force and on the drift angle which reduces to zero on the upwind and downwind headings. Therefore, the rate of change of airspeed (the slope of the line) is zero passing through the upwind and downwind headings. The rate of change of airspeed is greatest turning through the crosswind headings (090 and 270). (In relatively low-G albatross dynamic soaring, the effect of R_H is dominant giving the airspeed gain in the windward turn).

During the first and fourth quarters, making up the windward turn, the aero-forces cause the groundspeed to reduce and the airspeed reduces also. The rate of turn causes the headwind component to increase or tailwind to decrease (R_H is positive) which only reduces airspeed losses.

During the middle half of the circle, the leeward turn, the rate of turn causes the headwind component to reduce but this is offset by the effect of the wind-gradient to increase the headwind component. Groundspeed increases at such a rate that the airspeed increases as well but at a lesser rate. Thus there is a net gain of airspeed and groundspeed. These graphs use the derivative version (see the appendix) and therefore R_K and R_H are not calculated separately.

Drift

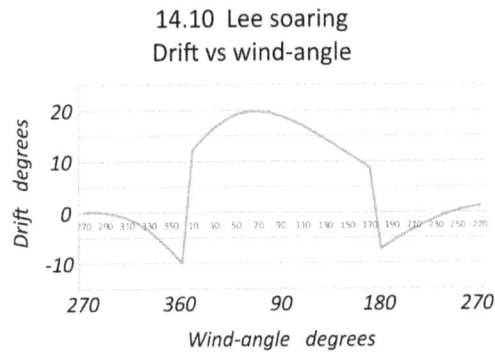

14.10 Lee soaring
Drift vs wind-angle

The drift angle is greatest in the middle of the leeward turn at the greatest height and the greatest wind. The effect of the drift angle is to enable the centripetal forces, which are making the aircraft turn, to also act to make the ground-speed increase. The kinks in the lines are the points at 360 and 180 degrees where the drift angle reduces to zero and increases again on the opposite side; that is, the drift angle flips from left to right or right to left. (Figure 14.10)

Drift is also approximately zero on wind-angle 270, at the lowest point of the circle, at the bottom of the wind-gradient where the wind is at a minimum and this will minimise the effect of R_K which is making the ground-speed reduce at this point.

Drift is either positive, on the same side as the angle of bank or negative, on the opposite side as the angle of bank.

Rate of change of airspeed

The sum of these effects gives the rate of change of airspeed R_V as shown in figure 14.11. That, in turn, leads to the actual airspeed as shown earlier. When R_V is greater than zero, airspeed is increasing and when R_V is less than zero, airspeed is decreasing. This is a reassuringly smooth curve despite the joggles in the drift curve.

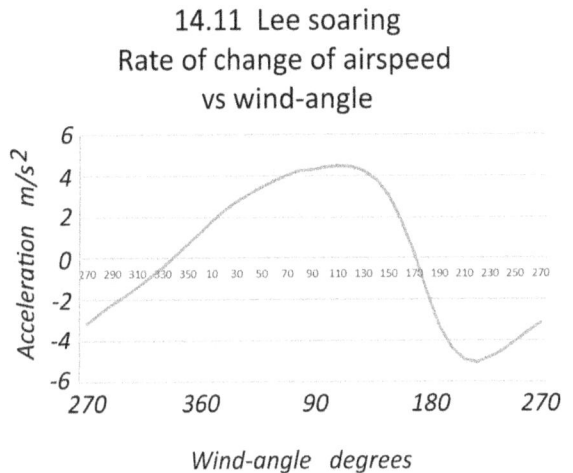

14.11 Lee soaring
Rate of change of airspeed
vs wind-angle

Momentum

The exchange of momentum and energy with the wind is essentially the same as with albatross dynamic soaring but with different combinations of drift and rate of turn. See figure 14.12 which shows a circling glider at the leeward and windward crosswind points. The wind is from the left. M is the momentum imparted to the air due to the centripetal force C. M_1 and M_2 are the components of momentum opposite to and in the same direction as the wind, respectively. The centripetal force is of course due to the resultant and not just the lift.

In the leeward turn, part of the momentum given to the air is opposite to the wind-velocity; therefore, the wind loses momentum as the aircraft gains momentum. In the windward turn the aircraft loses momentum and the wind gains momentum but overall, the aircraft gains more momentum than it loses and comes out with more speed. This is because the angle of drift in the leeward turn is greater at the top of the wind-gradient compared with the smaller angle of drift in the windward turn at the bottom of the wind-gradient where the wind is less. This means that the effect of the aerodynamic forces on the tangential acceleration of the aircraft are greater in the leeward turn than in the windward turn, and therefore the exchange of momentum is biased in favour of the aircraft.

14.12 Lee soaring
Transfer of momentum in leeward and windward turns

Plan view of glider in steeply banked cicling flight

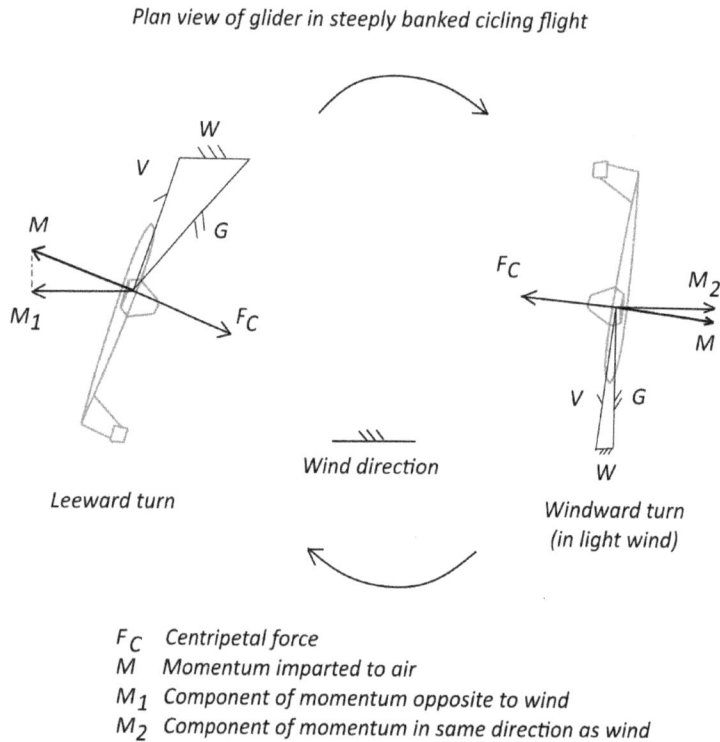

Wind direction

Leeward turn

Windward turn
(in light wind)

F_C Centripetal force
M Momentum imparted to air
M_1 Component of momentum opposite to wind
M_2 Component of momentum in same direction as wind

Interpretation

These results are difficult to interpret. The changing headwind component depends on both the change of height of the glider in the wind-gradient and on the change of direction relative to the wind direction. The biggest gain of airspeed is in the leeward turn where the headwind component is reducing ($\mathbf{R_H}$ is negative) but the effect of $\mathbf{R_K}$ is greatest ($\mathbf{R_K}$ is positive).

I take this to indicate that the principle effect on the airspeed of the glider is the acceleration due to the aerodynamic forces and the angle of drift. The airspeed does not increase simply because the wind increases with height. The effect of the wind-gradient is to increase the angle of drift in the leeward turn at the top of the climb. The effect of the artificially high-G leeward turn and the large drift-angle is to rapidly increase the ground-speed and thereby increase component **K**. The acceleration of airspeed is then the sum of the rate of change of component **K** and the rate of change of component **H**. $\mathbf{R_K}$ is positive and greater in magnitude than $\mathbf{R_H}$ which is negative and therefore $\mathbf{R_V}$ is positive. Therefore, airspeed increases in the leeward turn.

The difference between RC Dynamic soaring and albatross dynamic soaring

The maths in this RC dynamic soaring illustration are essentially the same as that used to describe the flight of the albatross according to the Windward Turn Theory but the geometry is different. The asymmetry of the diagrams is caused by the horizontal axis being intervals of wind-angle and not equal time.

The difference is that RC dynamic soaring uses a stationary hill and high-G turns in a wind-gradient to achieve high airspeeds while maintaining the position of the circle relative to the ground. The diagrams of bank and load-factor show the variation of bank from 32 to 68 degrees and of load-factor from 2 to 5G. At higher speeds the load-factor will have to increase to maintain the circle, requiring specially built, immensely strong gliders. To achieve the very high airspeeds exceeding 500 mph (225 m/s) in relatively small circles and stronger winds requires G-loading exceeding 100G!

This is quite different to classic albatross dynamic soaring in which the albatrosses do not fly circles but rather fly short segments of the windward and leeward turns, at low-G and reversing the direction of the turn between each segment. Albatrosses cannot tolerate the energy penalty associated with high-G flight. The albatrosses leeward turn is flown as a low-G wing-over with a large angle of bank but at a relatively small angle of attack. Their objective is to achieve great crosswind distance with minimum effort or a short upwind dash at slightly greater cost but with the prospect of a meal at the end.

The slingshot analogy

This is all rather more complicated than the wind gradient theory. Sure, the RC glider climbs and descends through the wind gradient and gains airspeed but the process of turning adds another layer of complexity.

The easiest way to picture what is going on is to imagine you have a rock tied to a string and you are whirling it around your head, like a sling-shot. (Figure 14.13) The tension in the string is like the horizontal resultant aerodynamic force acting on the glider in the leeward turn. Your hand moves in a circle while pulling on the string, which means there is an angle between the string and the radius from the rock to the centre of the circle described by your hand, which is analogous to the angle of drift. This allows the string to apply both a centripetal force, giving the rock a curved path, as well as a tangential force making the tangential speed of the rock increase. Which of course increases both actual speed and airspeed.

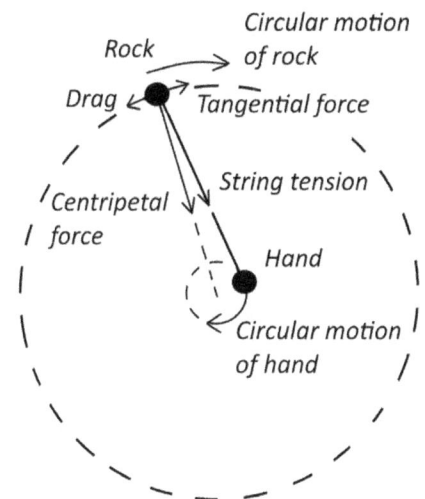

14.13 Slingshot analogy

In RC dynamic soaring, the effective centre of the circle is not in a fixed position but moves around in a circular motion, which is seen by comparing the ground and air plots in which the crosswind motion (as well as the up/downwind motion) is different in the two diagrams.

Now we understand how RC glider dynamic soaring works, the next chapter explains how the albatrosses use parts of a circle to dynamic soar upwind.

Chapter 15

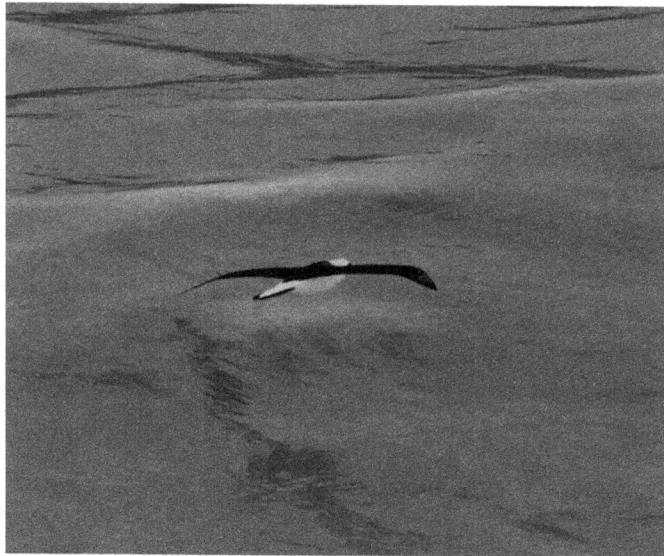

Albatross dynamic soaring
Upwind and downwind

Can an albatross use dynamic soaring to travel against the wind?

In cross-wind dynamic soaring at low-G, there is nearly always a large downwind drift angle. However, there is anecdotal evidence and there is data-logged evidence of upwind soaring, as you can see in chapter 9. Also, there has been heart rate monitoring of albatrosses suggesting higher metabolic rates on upwind legs compared with downwind or crosswind-soaring. This suggests that a different technique is being used to achieve upwind soaring and that, if the bird is not actually flapping, then one reason for this greater effort may be higher-G manoeuvring. The greater the load-factor the greater the effort on the part of the bird to keep its wings extended in the flight posture. Therefore, there is an energy budget which depends on the particular dynamic soaring technique and load-factor being used. The lowest load-factors are used in crosswind dynamic soaring and so it is not surprising that albatrosses spend a large part of their time flying in a crosswind direction. This will also give them the best chance of intercepting scent trails drifting downwind, which will lead them to their next meal. The most efficient energy budget would involve flying crosswind as much as possible to cover the greatest distance with the least energy loss, then making short upwind sprints when the scent trail of potential prey is detected.

The mathematical models used to simulate these manoeuvres depend on the shape of the turns and the load-factors being used. In crosswind dynamic soaring, the load-factor in a windward turn at a 10 degrees angle of bank is 1.015G. The leeward wing-over can be flown at a similar load-factor. However, in the upwind dynamic soaring manoeuvre described below, a 30 degrees banked turn requires a load-factor of 1.15G. A difference of 0.135G may not seem like much but it represents an increase of 13% in the effort needed to keep the wings extended, even if the albatrosses have a wing-joint locking mechanism. This must have an effect on the energy budget of an albatross during a foraging trip lasting many days.

The significance of the wind-gradient in cross-wind dynamic soaring is minimal. However, it may be that wind-gradients or wind-shears are used by albatrosses to achieve upwind and downwind dynamic soaring because the average heading is closer to the wind direction. Certainly the models work better with a wind-gradient included, although it is the wind that is the primary source of energy, not the wind-gradient. Also, the relatively high-G manoeuvring associated with upwind dynamic soaring may work in normal winds but will depend on how much effort the bird puts into it. It now seems likely that the crosswind, upwind and downwind dynamic soaring techniques all blend seamlessly together according to the bird's desire to fly in a particular direction. These manoeuvres all use essentially the same dynamics as described in chapter 13; the difference being the specific parameters such as wind-angle, angle-of-bank and load-factor which control the shape of the manoeuvre. There are two possible mechanisms involved.

The Surge

This is something that might happen when the bird is in a steeply-banked leeward turn passing through a crosswind heading. It gets a pulse of thrust, the surge, as it gains height and penetrates a shallow shear boundary into a stronger wind speed, perhaps in the lee of a breaking wave.

This is not the same as encountering a horizontal gust while in horizontal straight flight. It is more like the surge experienced by glider pilots flying into rising air causing the glider to briefly accelerate forward and upward before returning to equilibrium.

Imagine a glider, wings level, in normal straight 1G flight at the best lift/drag ratio angle of attack (3 to 4 degrees). The total aerodynamic force resolves into lift and drag respectively normal to and opposite to the direction of flight. If the glider flies into a <u>vertical</u> gust, the direction of the relative airflow changes, causing the angle of attack to increase close to the stalling angle of attack (about 15 degrees). The aerodynamic force is increased and because it is oriented to the new relative airflow, it becomes tilted forward of the aircraft vertical and then resolves into lift and thrust. The aircraft accelerates forward and upward. This accelerated motion reduces the angle of attack and the aircraft returns to a state of equilibrium, so that the effect can only be brief.

Now imagine the whole effect rotated through 90 degrees so that the bird is in a vertical bank and the wind gust is horizontal. (Figure 15.1). Consider the albatross dynamic soaring, making a regular leeward turn as a steeply banked wing-over but close to the lee-side of a steep swell or a breaking wave. In the lee of the wave, the wind may break away leaving a wind-shadow of relatively still-air in the trough and a marked horizontal shear boundary between the strong wind above and the still-air below. As the bird penetrates the horizontal shear boundary at a steep angle of bank, it encounters a sudden increase in horizontal wind-speed and a sudden increase in airspeed and angle-of-attack. The pulse of propulsive force is similar to the process of auto-rotation which drives windmills and spins helicopter rotors in power-off glides.

15.1 Dynamic soaring upwind
The Surge

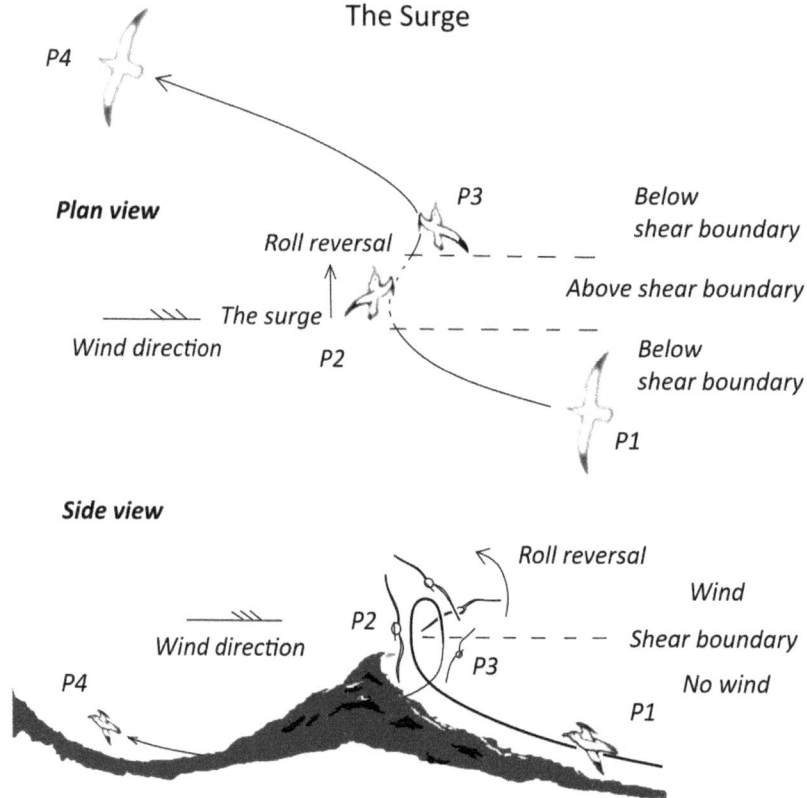

Plan view

P4

P3

Roll reversal

Below
shear boundary

Above shear boundary

The surge

Wind direction

P2

Below
shear boundary

P1

Side view

Roll reversal

Wind

P2

Wind direction

P3

Shear boundary

P4

No wind

P1

The lift and drag forces are components of the resultant. With the new relative airflow **V**, the resultant tilts forward of the glider vertical. (Figure 15.2). This will cause a component of aerodynamic force **T** to act momentarily as a pulse of thrust and cause the airspeed to increase. The other component **C** acts as the centripetal force, maintaining the turn. Because of the increased angle of attack, this will be inherently high-G and therefore will require more effort on the part of the bird to keep its wings extended. Having gained an increment of airspeed, the bird then reverts to normal angles of attack.

The bird can use the excess airspeed in any of three ways:

1 - It can gain height, or

2 - It can continue the turn and drop below the shear boundary in the downwind trough and gain distance downwind or crosswind, or

3 - It can reverse the direction of turn and drop below the shear boundary into the upwind trough and gain distance upwind, gliding as far as the next wave crest. If the swell is deep enough to create an air-flow separation, it may be possible to find relatively still-air in the troughs between wave crests, although the swells may be moving downwind. The increased load-factor due to the gust and the reversal of turn for upwind progress, will take more effort than rolling out of the turn into the downwind trough, hence the greater effort and metabolic rate during upwind dynamic soaring.

15.2 Aerodynamic forces during the surge

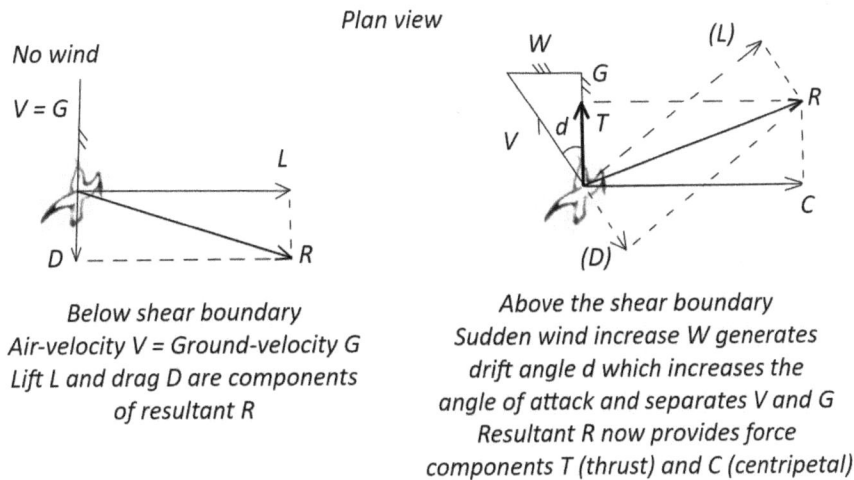

Plan view

Below shear boundary
Air-velocity V = Ground-velocity G
Lift L and drag D are components
of resultant R

Above the shear boundary
Sudden wind increase W generates
drift angle d which increases the
angle of attack and separates V and G
Resultant R now provides force
components T (thrust) and C (centripetal)

The Surge model depends upon there being a gust or a wind shear which will most likely occur in the lee of a breaking wave but cannot be guaranteed especially in light winds or in the absence of steep waves. Therefore, paradoxically it appeared that this particular dynamic soaring manoeuvre is only possible in strong winds. This gain of airspeed when penetrating a shear boundary is not the same as the wind-gradient theory of dynamic soaring.

A variation of the Windward Turn Theory

The second effect happens when the bird deliberately uses relatively high-G manoeuvring in the turns while climbing and descending in a wind-gradient. This again would require greater energy expenditure on the part of the bird to maintain its flight posture under high-G; although the G-loading is nothing like what the RC gliders experience.

The upwind dynamic soaring manoeuvre is shown in figure 15.3. From A to B is a partial windward turn but gaining height. The turn continues across the wind and becomes a leeward turn from B to C, still gaining height. At C the turn reverses at the top of the climb and becomes another windward turn descending to D and continues as a descending leeward turn from D to E. At the lowest point E, the turn reverses, to start again at A.

The basic mechanism here is the same as the Windward Turn Theory as explained in chapter 13, but with a different shape to the manoeuvre based on the GPS tracking data from chapter 9. The upwind dynamic soaring manoeuvre is a hybrid of the crosswind albatross method and the RC glider method and again the final result is obtained by an iterative process to find out what combination of bank angle and load-factor produces a net gain of airspeed and height at the end of the manoeuvre. Once again, the manoeuvre is modelled in an Excel spreadsheet, using the same set of equations as in the other sections and the output is in the form of graphs of each parameter.

15.3 Dynamic soaring upwind manoeuvre

A - B Windward turn
B - C Leeward turn
C - D Windward turn
D - E Leeward turn

Wind gradient

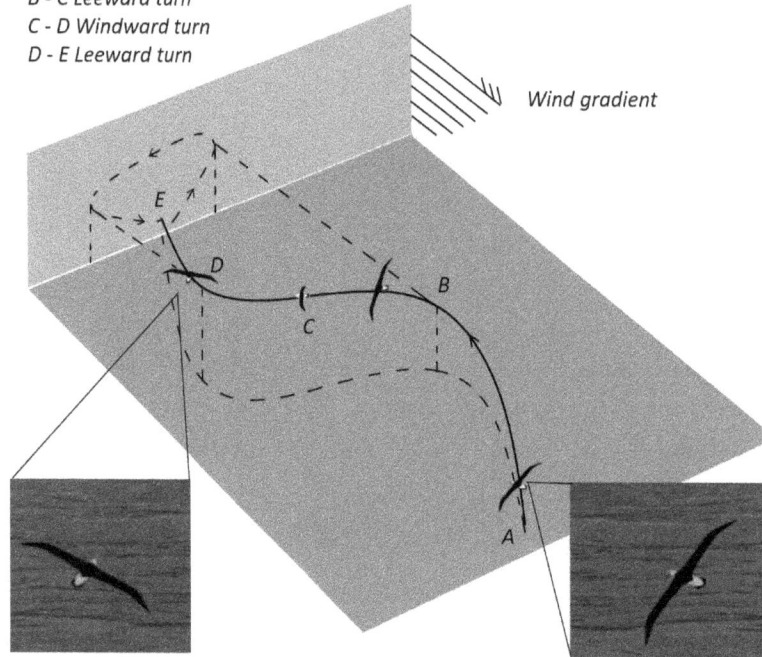

Upwind air-plot and ground-plot

Here is the output from the spread sheet. Figure 15.4 is a plan view of the path through the air with the wind from the top (North) and the bird flying from South to North. This time, the manoeuvre starts at a wind-angle of 60 degrees off the wind, at low level. The bird has just made a turn reversal from right to left bank. It turns upwind and across the wind at a minimum of about 30 degrees angle of bank to the left and a load-factor of about 1.15G. It pitches up, gaining height through the wind-gradient until, at the top of the climb at a wind-angle of 300 degrees, 60 degrees off the wind to the left, the bird reverses the angle of bank again to 30 degrees angle of bank to the right. It begins to descend, turning right through North, ending up at a wind angle of 060 degrees again and back at sea-level but with a slight excess of height and a couple of metres per second excess speed.

The ground plot (figure 15.5) is a plan view, showing the path over the ground, compressed by the Northerly wind. The drift is greatest in the middle of the manoeuvre, at the greatest height and wind speed.

15.5 Upwind dynamic soaring Ground plot
Plan view

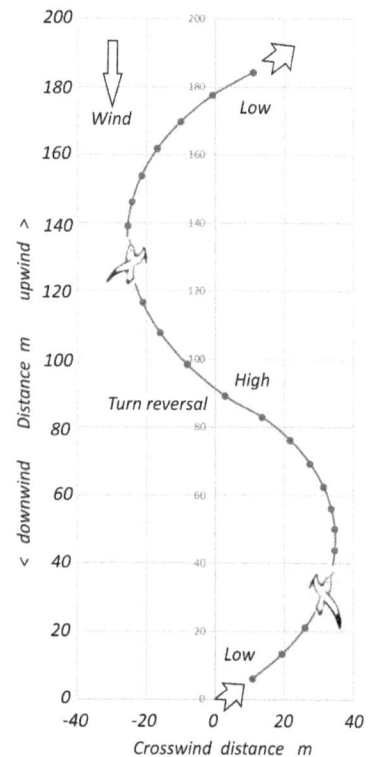

15.4 Upwind dynamic soaring Air-plot
Plan view

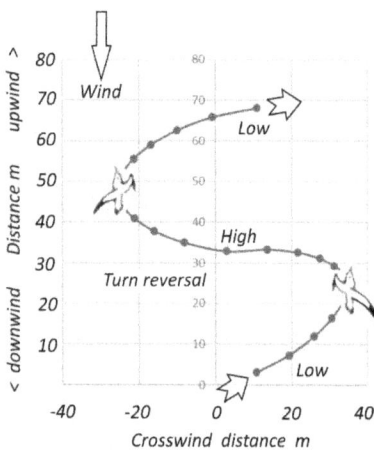

Wind gradient

The wind gradient is modelled so that the wind is proportional to log base 10 of the glider height. The wind-gradient against height is shown in figure 15.6. In the spreadsheet, the wind used depends on the glider height above an arbitrary datum.

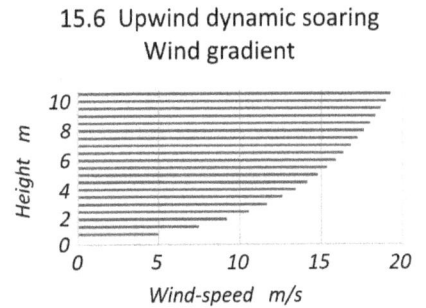

15.6 Upwind dynamic soaring
Wind gradient

Wind

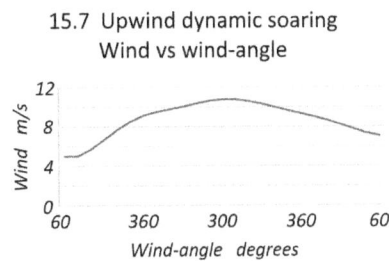

15.7 Upwind dynamic soaring
Wind vs wind-angle

The total wind (as opposed to the headwind component) encountered by the bird at each wind-angle is shown in figure 15.7. The wind varies from about 5 to 10m/s. Again, the greater wind at the end is due to finishing with a gain of height.

Height

Figure 15.8 shows bird height against wind-angle, starting at 1m. There is a reversal of bank-angle at the highest point at about 2.5m and ending with a slight gain of height to about 1.5m.

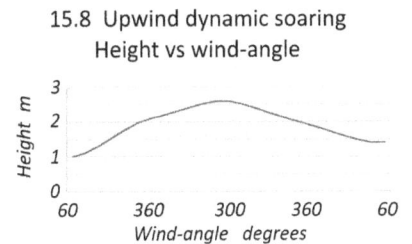

15.8 Upwind dynamic soaring
Height vs wind-angle

Airspeed and groundspeed

Airspeed and groundspeed are shown in figure 15.9. Starting at wind-angle 60 degrees, this is a windward turn. Groundspeed and airspeed both reduce as the bird gains height. After turning through North, wind-angle 360 degrees, the left turn continues as a leeward turn and groundspeed increases but airspeed increases as well due to the increasing headwind component due to the wind-gradient. The roll-reversal occurs at the mid-point, at the highest point, on a wind-angle of about 300 degrees, into another partial windward turn to the right and now the speeds start to decrease. Finally, after turning through North again, this turn becomes a leeward turn in the fourth quarter and the both speeds increase as height reduces and the wind reduces with height, ending up back on a wind-angle of 060 but with a slight excess of both groundspeed, airspeed and height.

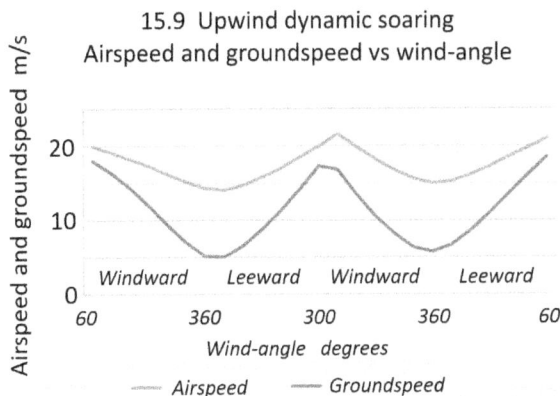

15.9 Upwind dynamic soaring
Airspeed and groundspeed vs wind-angle

This gain of airspeed will not be achieved simply by climbing and descending upwind without turning or by turning without climbing and descending. It requires a minimum angle of bank of at least 30 to 40 degrees which increases the load-factor and therefore the effort on the part of the albatross, by about 15 to 39%, compared to crosswind dynamic soaring. It also requires the wind-gradient to help generate the rate of change of airspeed but not exactly in the same way as the wind gradient theory supposes. The net result is a slight overall gain of airspeed. Remember, this is the result of applying the aerodynamic forces and the rate of change of headwind component to calculate the rate of change of airspeed and ground-speed during turning flight.

This result is similar to the result in the RC gliders lee-soaring section. The gain of airspeed and groundspeed takes place in the leeward turn sections because of the dominance of the groundspeed acceleration (R_K) compared to the rate of change of the headwind component (R_H).

For comparison, figure 15.10 is the airspeed and groundspeed in the RC gliders circular lee-soaring model again. Here, there are no turn reversals in the circular flight path, although the drift-angle flips between left and right drift at wind-angles 360 and 180. Both diagrams show a gain of speed in the leeward sections and a slight gain of airspeed at the end.

15.10 Lee soaring
Airspeed and groundspeed vs wind-angle

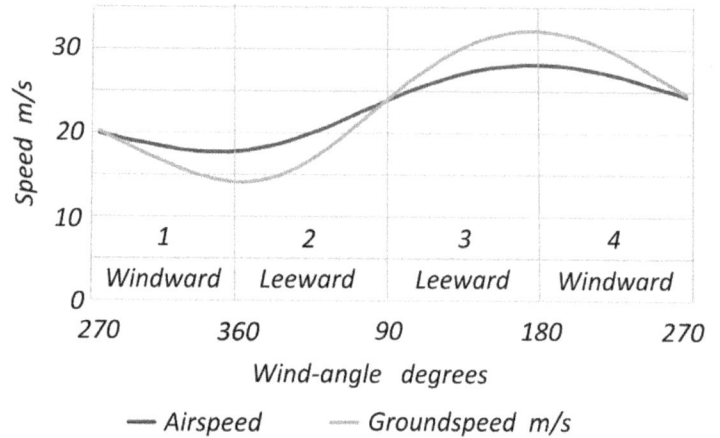

Dynamic soaring downwind

15.11 Lee soaring
Circular dynamic soaring

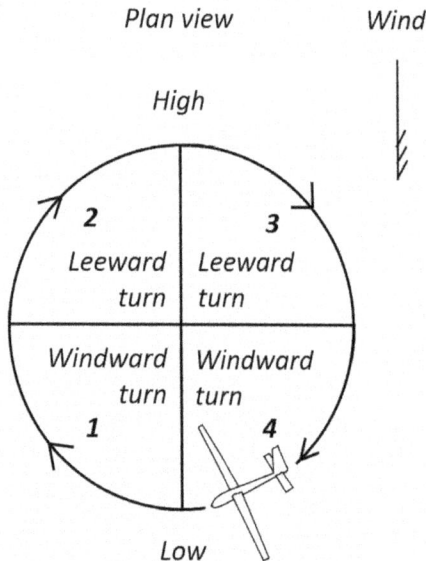

If the albatrosses can soar crosswind and upwind, can they soar downwind? You might think obviously yes; if crosswind dynamic soaring inevitably leads to down-wind drift, due to the drift angle, then surely they can dynamic soar directly downwind, albeit in the usual undulating fashion. Well, maybe not. In the Laysan albatross foraging data in chapter 9 there are many examples of crosswind and upwind dynamic soaring but nowhere is there any direct downwind dynamic soaring. There is one example of a kind of random circling and drifting downwind *at the speed of the wind* and elsewhere a distinctive pattern of zig-zagging downwind at an average downwind speed *greater than the wind*. We will look at that next.

Figure 15.11 shows in plan-view, the circular RC gliders manoeuvre which produces the variation of speeds shown in figure 15.10. The manoeuvre starts at the lowest point on a crosswind heading and turns right, upwind, climbing through the wind-gradient. The highest point is at the upwind apex and the turn continues downwind descending back through the wind-gradient to the starting point. The quadrants are numbered and labelled as windward turns or leeward turns corresponding to figure 15.10.

Next, in figure 15.12, are theoretical diagrams of the upwind and downwind dynamic soaring manoeuvres. We can see that the upwind manoeuvre, on the left, starting at the bottom, comprises only the sector 1 and sector 2 quadrants, with a turn reversal in the middle at the highest point. Whereas, the downwind manoeuvre on the right, starting at the top, comprises only sector 3 and 4 quadrants, again with a turn reversal in the middle but this time at the lowest point.

Referring back to Figure 15.10 we can see that sectors 1 and 2 yield a net gain of airspeed confirming that dynamic soaring upwind using sectors 1 and 2 is feasable; whereas sectors 3 and 4 give a net loss of airspeed so dynamic soaring directly downwind using sectors 3 and 4 in this fashion is not possible.

Bear in mind that this involves climbing and descending through a wind gradient, as indicated by the high and low labels and also a greater angle of bank and load factor with consequently greater effort on the part of the bird, compared with crosswind dynamic soaring.

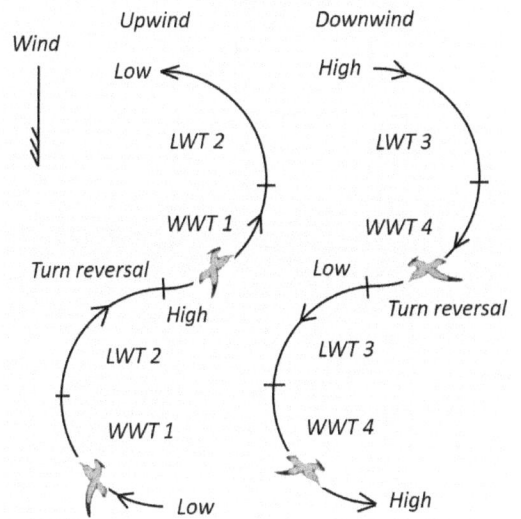
15.12 Upwind and downwind possibilities

Gybing downwind

What about the zig-zagging downwind, seen in the GPS data? It is possible to replicate this and get a result for downwind dynamic soaring by taking the classic crosswind dynamic soaring shown in chapter 13, figure 13.10, reducing the amplitude of the turns and moving the high and low spots. But it is marginal.

Downwind air-plot

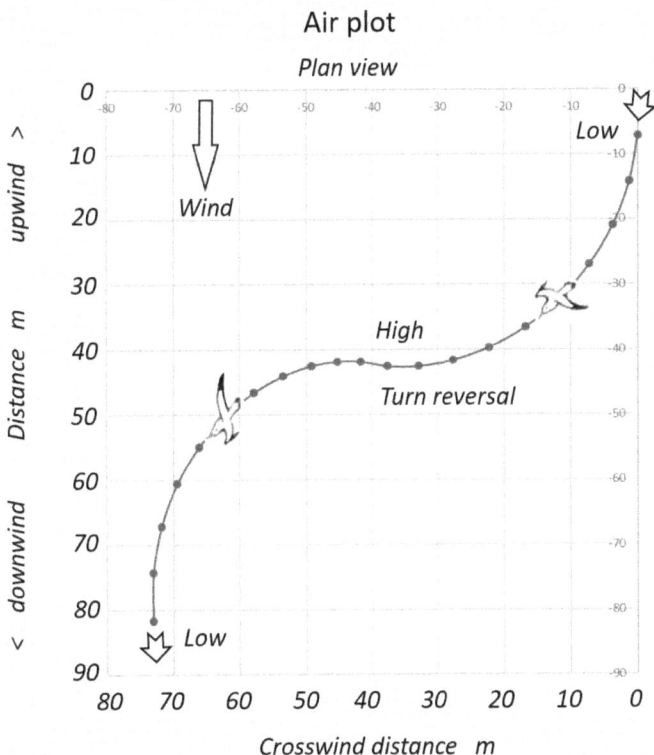
15.13 Downwind dynamic soaring Air plot

The dynamic soaring manoeuvre is flown at intermediate G-loading with a bank angle of about 45 degrees and a load-factor of 1.41G. This results in an average ground track at about 140 degrees to the wind, as in figure 15.13 which is a plan view of the air plot with the bird flying towards the South-West with a North wind.

Although the bird begins and ends the manoeuvre close to the downwind heading, the overall track-made-good is about 40 degrees off the directly-downwind track; in other words, at about 140 degrees of wind-angle. The first half of the manoeuvre is a partial windward turn, gaining height and losing airspeed The reversal of bank is at the high spot. The second half is a partial leeward turn losing height and gaining speed.

Downwind ground-plot

The next diagram (figure 15.14) shows the ground plot, stretched downwind by the tailwind and showing the actual track-made-good towards the South West. Of course, this can be followed by repeating as a right-left manoeuvre or reversed with a left-right pair of turns giving a more Southerly track-made-good.

15.14 Downwind dynamic soaring Ground plot
Plan view

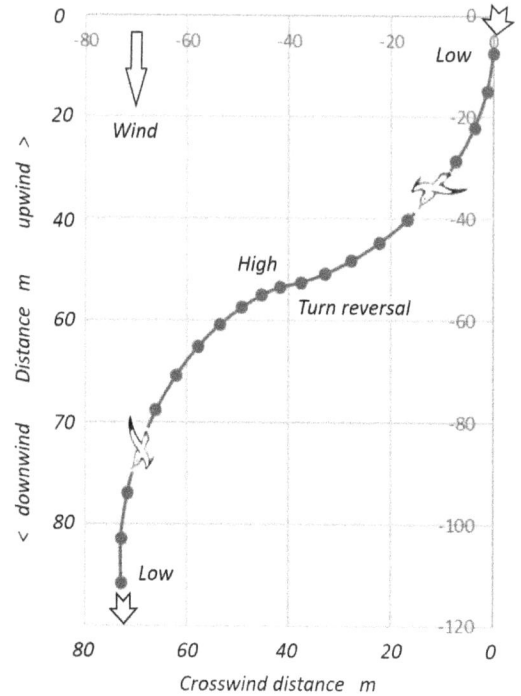

Airspeed acceleration

Figure 15.15 shows the rate of change of airspeed during the downwind manoeuvre. Airspeed reduces at an increasing negative rate, during the first half of the manoeuvre as height is gained then, with the roll-reversal, the airspeed increases but again at a reducing rate.

15.15 Downwind dynamic soaring Rate of change of airspeed

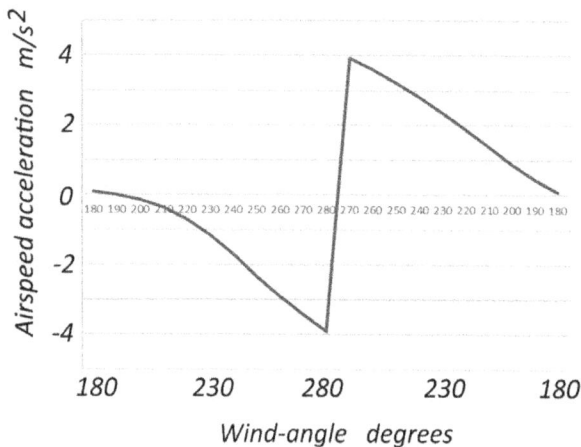

Airspeed and groundspeed

Figure 15.16 is the airspeed and groundspeed which both reduce to a minimum at the maximum height at the mid-point and then increase with a net gain of 0.5m/s airspeed.

The difference between the airspeed and groundspeed at the end of the manoeuvre is greater than at the beginning. This is because the bird has gained height at the end and the wind strength is therefore greater at the greater height within the wind-gradient.

15.16 Downwind dynamic soaring Airspeed and groundspeed vs wind-angle

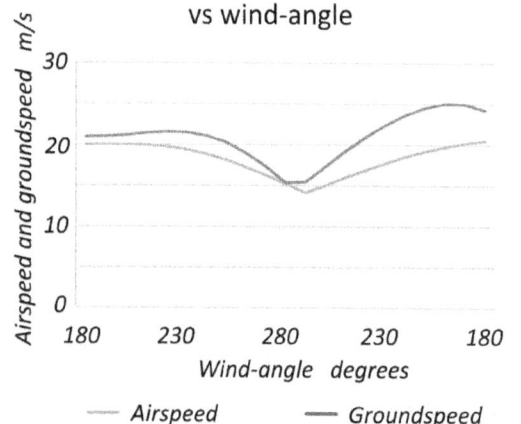

Height

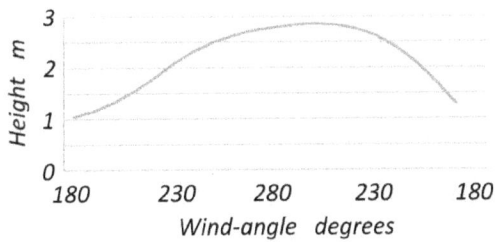

15.17 Downwind dynamic soaring
Height vs wind-angle

The height result is shown in figure 15.17, greatest in the middle and with a slight height gain of about 0.5m at the end.

Downwind dynamic soaring in nature

The manoeuvre, as performed by a real albatross, is seen in figure 15.18 which is a close-up of a section of the actual albatross tracking data from chapter 9, the section at about 897km North of departure.

The wind is from the south-east at 4 m/s. The bird's average groundspeed is about 17 m/s along its actual track while it achieves a mean track-made-good towards the North-West. The straight line distance is 3000 m travelled in 5 minutes at 10 m/s, which is faster than the wind. The bird's track is seen to zig-zag downwind as it links 4 or 5 dynamic soaring manoeuvres on one side of the wind and then turns across the wind and does another 4 or 5 DS manoeuvres on the other side and then switches back again. This is a bit like gybing downwind in a sailboat and achieving a downwind speed greater than the wind. Zig-zagging is apparently a faster, possibly more efficient way of travelling downwind compared with drifting with the wind but it is with relatively high-G manoeuvring.

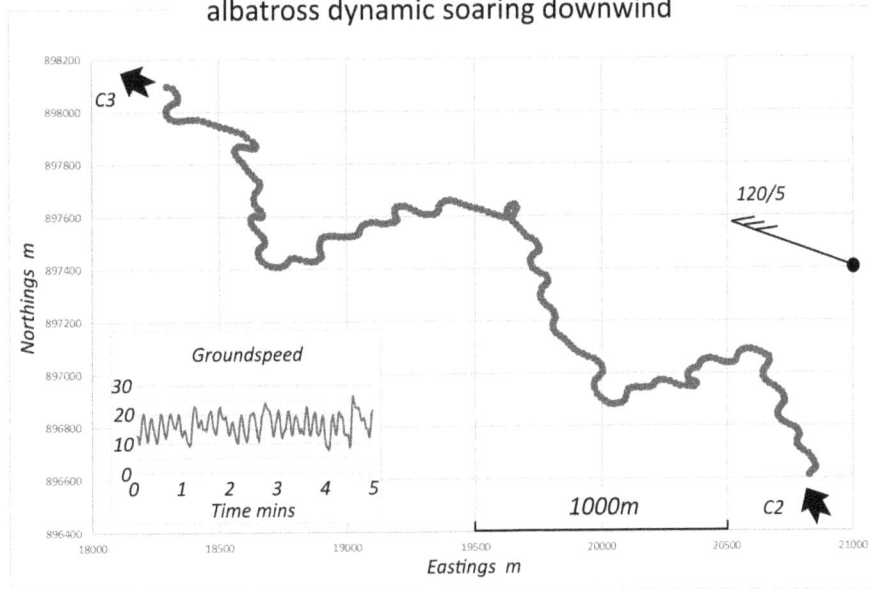

15.18 GPS plot of 5 mins of albatross dynamic soaring downwind

Other sections of the tracking data show the albatross moving downwind but apparently meandering at random and drifting downwind at the speed of the wind (Figure 9.15).

Epilogue

That concludes the story of dynamic soaring. It is more complicated than the Victorians imagined but then so is flying in general. Man has had the means and materials to achieve gliding flight for at least two millennia – at its simplest it is only carpentry and needlework. And yet it took many generations to figure out the secrets of flight; not only how to build a glider but also how to fly it. The secrets of the albatrosses remained hidden until the era of satellite navigation and the application of some mathematics and piloting technique.

Appendix 1 contains the scary maths which are used to produce the diagrams of dynamic soaring flight, which some viewers might find distressing. Just kidding! It really is no more than high-school maths. Dynamic soaring is not difficult like string-theory – it is more like tangled string which just needs to be un-picked!

Appendix 1

A mathematical model of dynamic soaring

Developing the model

Dynamic soaring depends on there being a single basic mechanism, controlled primarily by the angle of bank and load factor, which will enable the bird to maintain airspeed and height throughout the windward and leeward turns. That mechanism is the tendency of the airspeed to change due to two effects. Firstly, the changing headwind component caused by the changing wind-angle, that is the rate of turn. Secondly the opposite acceleration of ground-speed caused by the aerodynamic forces acting on the bird. The same basic mechanism must also explain upwind and downwind dynamic soaring and RC glider dynamic soaring. The rest of this chapter is an explanation of the maths used to produce the graphical results seen in the previous chapters.

R_W Rate of change of the headwind component H versus the wind angle y

Headwind component **H** is a function of trigonometry and simply depends on the aircraft heading relative to the wind, the wind-angle **y**. Given wind-speed **W** there is a headwind component **H** corresponding to each wind-angle **y**.

$$H = W \cdot \cos y$$

Taking a given interval between wind-angles, (say 1 degree), then in a turn between each pair of wind angles there is a corresponding change of headwind component. Therefore, we get a rate of change of headwind component with respect to the wind-angle.

The rate of change of head-wind component per degree of wind-angle:

$$R_W = [(W \cdot \cos(y+1)) - (W \cdot \cos y)]/1$$

$$R_W = W \cdot (\cos(y+1) - \cos y) \tag{1}$$

Where **W** is the wind and **y** is the wind angle.

Note that R_W (m/s per deg), does not depend on the rate of change of **y** with time. It simply depends on the particular wind-angle and the increment.

R_H Rate of change of headwind component with time

The rate of change of the headwind component R_H in units of m / sec^2 is the product of the rate of change R_W of the headwind component with respect to the wind-angle and the rate of change R_y of the wind-angle over time, which is the same as the aircraft rate of turn:

$$R_H = R_W \cdot R_y \tag{2}$$

Looking at the units here we get:

$$m / sec^2 = (m / (sec . deg)) \times (deg / sec)$$

The Drag Force F_D

The drag force depends on the lift force divided by the lift/drag ratio L_d. The lift force in level flight is equal to the weight (m.g) at 1G. In a level turn the lift force is then multiplied by the load factor L_f which is inversely proportional to the cosine of the angle of bank **x**. Note that, for aircraft, the lift/drag ratio varies with the angle-of-attack which also changes the load factor but in birds the relationship is unknown, so I will assume constant L/D ratio over the small range of airspeeds and angles-of-attack used in dynamic soaring:

$$F_D = (m . g . L_f) / L_d \qquad (3)$$

The Centripetal Force F_C

The centripetal force F_C in level flight (see figure 16.1) depends on the lift (weight) multiplied by the tan of the angle-of-bank or, in a wing-over, by the sine of the angle-of-bank times the load-factor:

$$F_C = m . g . L_f . \sin x \qquad (4)$$

Angle of bank and radius of turn in level flight

The angle of bank is measured from the vertical. The centripetal acceleration it provides is horizontal. See figure 16.1 in which a level turn is depicted where the vertical component of lift F_V is equal to the weight.

Vertical component of lift

$$F_V = Weight = m . g$$
$$Lift = m . g . L_f$$
$$= m . g / \cos x$$

Centripetal acceleration

$$F_c = m . g . \sin x / \cos x$$
$$= m .g . \tan x$$
$$F_c = m . a$$
$$\cancel{m}. a = \cancel{m} . g . \tan x$$

Radius of turn

$$a = v^2 / r$$
$$v^2 / r = g . \tan x$$
$$r = v^2 / g . \tan x \qquad (5)$$

x is angle of bank
a is centripetal acceleration m/s^2
v is tangential air-speed m/s
r is radius of turn m
m is mass
g is acceleration of gravity

16.1 Load factor

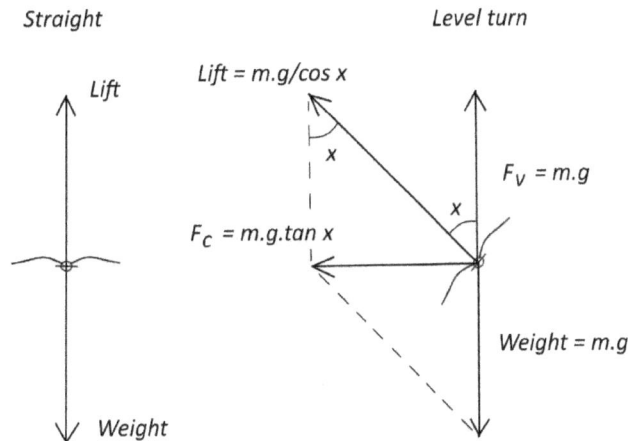

Straight

Level turn

Lift

Lift = m.g/cos x

x

$F_v = m.g$

$F_c = m.g.tan\,x$

x

Weight = m.g

Weight

Rate of turn R_y

The rate of turn R_y is a function of tangential speed **v** and radius **r**.

$$R_y = 360 \cdot v / C$$

$$R_y = 360 \cdot v / (2 \cdot pi \cdot r) \tag{6}$$

$$R_y = 360 \cdot v \cdot g \cdot Tan\,x / (2 \cdot pi \cdot v^2) \tag{6,5}$$

$$R_y = 180 \cdot g \cdot Tan\,x / (pi \cdot v) \tag{7}$$

R_y is rate of turn deg/sec

C is circumference of circle radius r

R_y depends on **x** and **v**, assuming a level turn in which the vertical component of lift is equal to the weight. In this equation **v** is the tangential speed which, in still air, would be airspeed equal to groundspeed. However, in a wind with an angle of drift which speed do we use, airspeed or groundspeed? The best result is given by using component **K**, that is groundspeed times the cosine of the angle of drift. This is a debateable point but has been discovered by iteration.

In the leeward turn, which is a wing-over and not level flight, we need to nominate the load-factor.

Load factor L_f

Load factor is the ratio of lift to aircraft weight. In a level turn the vertical component of lift must remain equal to weight and to achieve this the actual lift is increased by increasing the angle-of-attack. The load factor is equal to 1/ cosine of the angle-of-bank. For example, in a 30 degree banked level turn the load factor is 1.15G.

In a wing-over at any given angle-of-bank, the load-factor can be whatever we want but the vertical component of lift will not necessarily be equal to the weight. In this case we can introduce the load-factor to equation (7) by saying that:

$$tan\,x = sin\,x / cos\,x$$

and

$$L_f = 1/cos\,x$$

Therefore

$$\tan x = L_f . \sin x \qquad (8)$$

Substituting in equation (7) the Rate of turn is

$$R_y = 180 . g . L_f . \sin x / (pi . v) \qquad (7,8)\ (9)$$

Note that when this equation is used in the windward turn we nominate the load-factor which will give level flight. In the leeward turn we use an arbitrary 1G in the wing-over combined with suitable angles of climb and descent. And we use **K** instead of **v**.

Horizontal resultant $\mathbf{F_R}$ and components

Figure 16.2 is a plan view of a windward turn with the wind coming from the left of the diagram and the bird is flying in a banked left turn. The horizontal component of lift $\mathbf{F_C}$ is normal to the direction of the air velocity and combines with the drag $\mathbf{F_D}$ to create the horizontal resultant $\mathbf{F_R}$.

16.2 Force components
relative to air and ground-velocities
for a glider in a windward turn

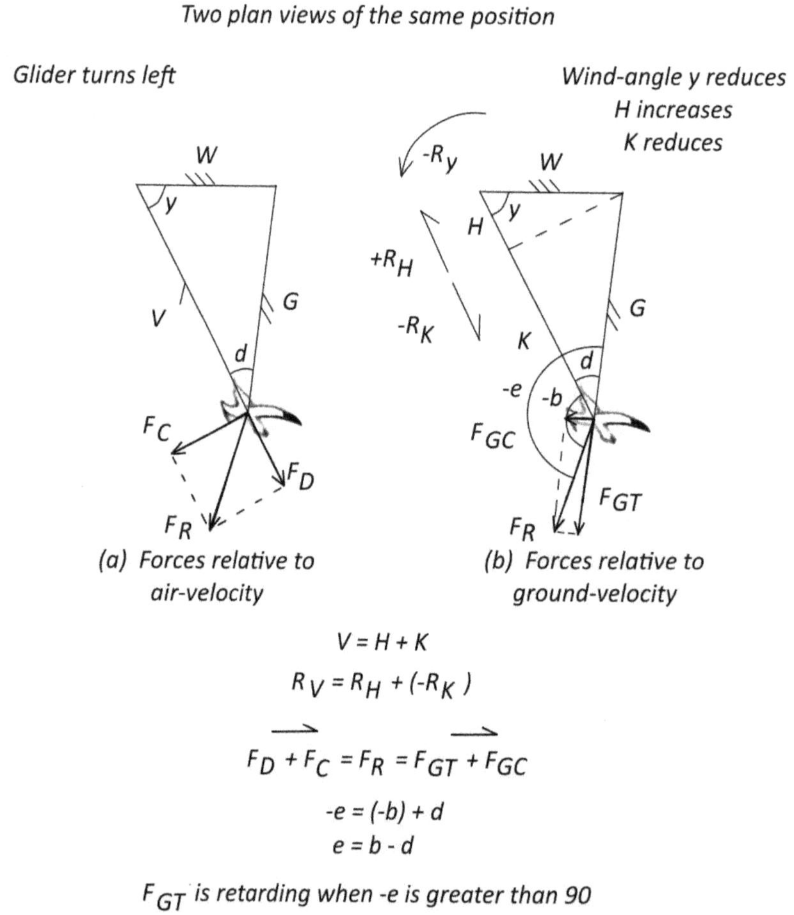
Two plan views of the same position
Glider turns left
Wind-angle y reduces
H increases
K reduces
W
-Ry
y
V
G
d
+RH
-RK
H
K
-e
-b
FC
FD
FR
FGC
FGT
(a) Forces relative to air-velocity
(b) Forces relative to ground-velocity
V = H + K
RV = RH + (-RK)
FD + FC = FR = FGT + FGC
-e = (-b) + d
e = b - d
FGT is retarding when -e is greater than 90

We can calculate $\mathbf{F_R}$ using $\mathbf{F_D}$ and $\mathbf{F_C}$ and Pythagoras which means that $\mathbf{F_R}$ is always positive:

$$F_R = \text{sqrt} (F_D{}^2 + F_C{}^2)$$

(The ultimate effect of $\mathbf{F_R}$ will depend on the sign of angle **e** which will change with the angle of bank. The sign of the angle of bank will be positive in a leeward turn and negative in a windward turn).

Forces and angles

Force $\mathbf{F_R}$ is not exactly in line with the ground velocity, and can be resolved into components $\mathbf{F_{GT}}$ the ground tangential force and $\mathbf{F_{GC}}$ the ground centripetal force. It can be seen that, when the wind is a large proportion of the airspeed, the drift angle **d** is also large. In albatross dynamic soaring, the drift angle is on the same side in both windward and leeward turns and is taken to have a positive sign. In the windward turn, the sign of the angle of bank is negative, because it is on the opposite side to the angle of drift; it then changes to positive in the leeward turn, because it is then on the same side as the angle of drift. This is important because it ultimately determines whether the tangential forces are positive or negative, that is, propulsive or retarding.

Forces are positive in the direction of flight, that is the direction of the air-velocity, therefore force $\mathbf{F_D}$, the drag force, has a negative sign. $\mathbf{F_C}$ the centripetal force also has a negative sign because of the negative angle of bank in the windward turn (opposite side to the angle of drift). The angle of drift **d** is the angle between the air-velocity and the ground velocity. Angle **d** is also the angle between the accelerations of air-velocity and ground-velocity. The angle **b** is the angle between $\mathbf{F_R}$ and the direction of the air-velocity but it depends on the ratio of $\mathbf{F_C}$ and $\mathbf{F_D}$. Angle **e** is the angle between $\mathbf{F_R}$ and the direction of the ground-velocity.

Force components relative to the ground-velocity

In the windward turn a tangential force component $\mathbf{F_{GT}}$ acts opposite to the direction of the ground-velocity:

$$F_{GT} = F_R . \cos e \qquad (10)$$

Note that angle **e** is not the internal angle between $\mathbf{F_R}$ and $\mathbf{F_{GT}}$ but is the external angle measured from the forward direction. $\mathbf{F_R}$ is positive and **e** has a large negative value which ensures that $\mathbf{F_{GT}}$ is negative and therefore a retarding force.

The centripetal component $\mathbf{F_{GC}}$ provides the centripetal acceleration which creates the curved path relative to the ground:

$$F_{GC} = F_R . \sin e \qquad (11)$$

Once again $\mathbf{F_{GC}}$ will have a negative value because e is negative (opposite to the angle of drift in the windward turn) and greater than 90 degrees.

Angles b, d & e

Angle **b** is the angle between $\mathbf{F_R}$ and the glider heading:

$$\sin b = F_C / F_R$$

$$(\sin b = \sin (180-b)$$

Angle **e** is the angle between $\mathbf{F_R}$ and the ground track (the direction of the tangential acceleration)

The tangential acceleration of the ground-velocity is:

$$A_{GT} = F_R \cdot \cos e \, / \, m \tag{12}$$

The centripetal acceleration of the ground-velocity is:

$$A_{GC} = F_R \cdot \sin e \, / \, m \tag{13}$$

Sorting out the positives and negatives

Angles **b**, **d** and **e** are illustrated in figure 16.2 for the windward turn and figure 16.3 for the leeward turn. By inspection, in the windward turn angle **e = b + d**. Angle **d** is the angle of drift and has a positive sign. Angles **b** and **e** are on the opposite side and are therefore negative:

$$(-e) = (-b) + (+d)$$

$$e = b - d$$

In the leeward turn in figure 16.3, by inspection, **e = b - d** but all the angles are on the same side and have the same sign (positive) therefore:

$$e = (+b) - (+d)$$

$$e = b - d \tag{14}$$

16.3 Force components for a glider in a leeward turn with a large angle of bank

Two plan views of the same position

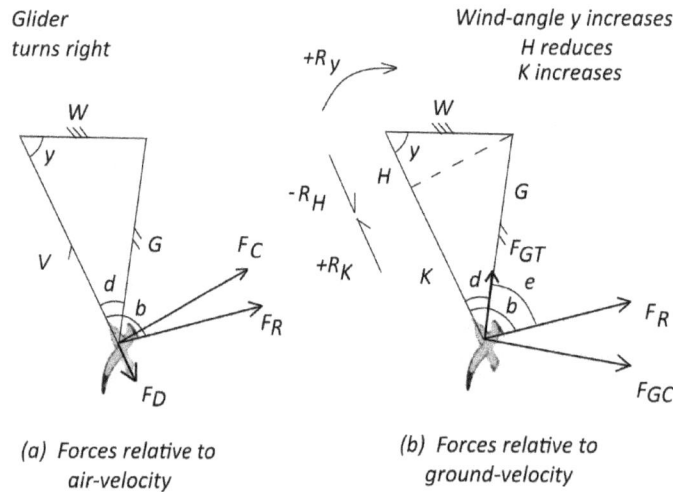

(a) Forces relative to air-velocity

(b) Forces relative to ground-velocity

$$V = H + K$$

$$R_V = (-R_H) + R_K$$

$$\overrightarrow{F_D} + F_C = F_R = \overrightarrow{F_{GT}} + F_{GC}$$

$$e = b - d$$

F_{GT} is propulsive when angle e is less than 90 degrees

Thus, in the calculations, we use the same formula **e = b - d** in both the windward and leeward turns. Angle **e** then has a large negative value in the windward turn and a small positive value in the leeward turn. This results in ground tangential force $\mathbf{F_{GT}}$ and acceleration $\mathbf{A_{GT}}$ being retarding in the windward turn and propulsive in the leeward turn. The sign of the drag force $\mathbf{F_D}$ is negative, opposite to the direction of flight. The value of centripetal force $\mathbf{F_C}$ can be negative or positive depending on the angle of bank which is measured left or right from the vertical and is negative for a windward turn and positive for a leeward turn. The angles **b**, **d** and **e** are merely artefacts of the model. All that the albatross experiences are the forces.

(Note that in a straight level glide, when the angle of bank is zero, then $\mathbf{F_C}$ is zero. $\mathbf{F_R}$ is the then same as $\mathbf{F_D}$ and angle **b** is 180 degrees. Angle **e** is **180 - d**, large positive angle and **cos 180 - d** is a large negative and therefore $\mathbf{F_{GT}}$ is retarding).

This seems rather complicated and if the albatrosses did not exist, you might think that it is all a bit far-fetched. However, it is necessary to treat all the forces equally and account for the acceleration of ground speed. This treatment of the force components is the only way to produce a realistic model of albatross flight.

Components of airspeed and ground-speed

So far we have seen how the aerodynamic forces cause the ground-speed to decrease in the windward turn and increase in the leeward turn. Does this have an effect on the airspeed?

There is a component of the ground speed **K = G . cos d** and a component of the wind-speed **H = W . cos y** Both are parallel with the air-velocity which is the sum of the two components. (Figure 16.3)

$$\text{Airspeed } V = K + H.$$

The rate of change of airspeed $\mathbf{R_V}$ is the sum of the rate of change $\mathbf{R_H}$ of the headwind component **H** and the rate of change $\mathbf{R_K}$ of the ground-speed component **K**.

$$R_V = R_H + R_K$$

In order for the airspeed to be constant, the ground-speed component **K** must reduce and the wind component **H** must increase at the same rate $\mathbf{R_H}$

$$R_H = - R_K$$

But force $\mathbf{F_D}$ does not directly affect component **H**, it only affects component **K** by causing a tangential acceleration of the aircraft.

To understand this, compare with straight and level flight in a uniform wind where the headwind component **H** is constant because the aircraft is not turning. Therefore, an unbalanced drag force $\mathbf{F_D}$ would cause the airspeed **V** to reduce but in reality it is only making component **K** reduce. Therefore, we use component **K** in the acceleration calculations and not airspeed **V**. This is arguable but you can run around in circles trying to justify which to use. It is one of the items that was settled by iteration. **K** gives a better result than **V**.

Acceleration of component K

The tangential acceleration $\mathbf{R_K}$ of the component **K** will be the acceleration $\mathbf{A_{GT}}$ of groundspeed **G** reduced by the cosine of the angle of drift. The ground tangential acceleration is caused by the ground tangential force $\mathbf{F_{GT}}$ which in turn, is a component of force $\mathbf{F_R}$.

$$R_K = A_{GT} . \cos d$$
$$= F_{GT} . \cos d / m$$
$$R_K = F_R . \cos e . \cos d / m \qquad (14)$$

As mentioned before, the calculation gives angle **e** a large negative value in the windward turn which makes $\mathbf{R_K}$ a retarding tangential acceleration.

$\mathbf{R_y}$ Rate of Turn

The rate of turn depends on the centripetal force $\mathbf{F_C}$. This gives a good result in the spreadsheet because $\mathbf{F_C}$ is the centripetal component of $\mathbf{F_R}$ in the air frame of reference. Any other treatment of the centripetal forces will reduce the rate of turn and reduce the rate of change of the headwind component and the rate of increase of airspeed in the windward turn.

Referring to equation (8) the rate of turn is:

$$R_y = 180 \cdot g \cdot L_f \cdot \sin x / pi \cdot v \qquad (15)$$

The term $\mathbf{g} \cdot \mathbf{L_f} \sin \mathbf{x}$ is the centripetal acceleration caused by the angle-of-bank in units m/s^2 In the spreadsheet this is substituted with centripetal acceleration $\mathbf{F_C} / \mathbf{m}$ which depends on angle-of-bank $\mathbf{x}$. The velocity term $\mathbf{v}$ is substituted with component $\mathbf{K}$

$$R_y = F_C \cdot 180 / m \cdot pi \cdot K \qquad (16)$$

$\mathbf{Ry}$ can be used to calculate the time interval between successive increments of wind-angle:

$$T = \Delta y / R_y$$

Creating a spreadsheet to optimise angle of bank

Now, we can plot values of $\mathbf{R_K}$, $\mathbf{R_H}$ and $\mathbf{R_V}$ against a range of angles of bank to produce the optimum angle of bank result seen in Chapter 13 repeated here as figure 16.4. This gives small angles of bank which correlates quite well with the small angles of bank observed in film of albatross' windward turns. The nominated values of airspeed and wind-speed can then be modified to see, for example, what the minimum wind speed needs to be.

This answers the question we set ourselves: that airspeed can be constant or even increase slightly during a windward turn, even though there is an unbalanced drag load and the ground speed is reducing. This is somewhat counter-intuitive but fortunately, it demonstrates the viability of the Windward Turn Theory.

16.4 Glider windward turn
Acceleration vs angle of bank

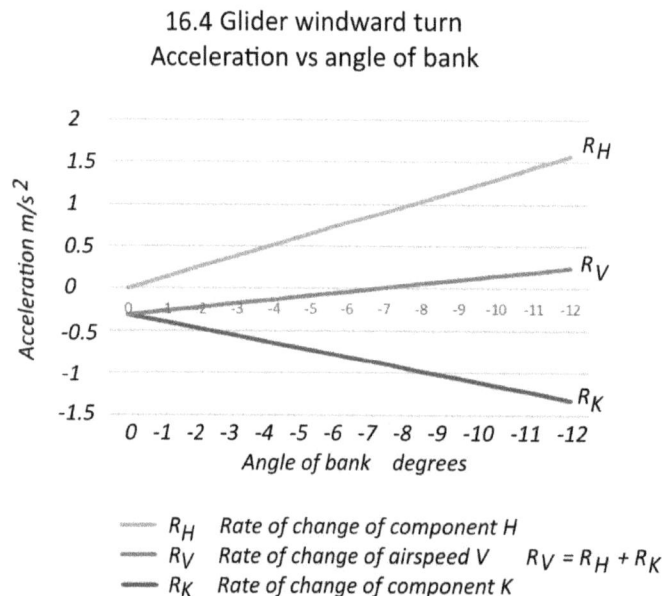

	R_H	Rate of change of component H
	R_V	Rate of change of airspeed V $\quad R_V = R_H + R_K$
	R_K	Rate of change of component K

The full list of rows for this spreadsheet is as follows:

Mass m for example 10 kg

L/d ratio L_d for example 20

Angle of bank x a range of angles from 0 to -12

Load factor $L_f = 1/\cos x$

Wind angle y a single wind-angle of 90 degrees for a snap shot

Wind speed W uniform for example 10 m/s

Airspeed V not assuming constant airspeed but just a representative value for example 20 m/s

Groundspeed $G = \text{sqrt} [(V^2 + W^2) - 2 \, V.W \cos y]$

Drift $d = \cos^{-1} [(V^2 + G^2 - W^2) / (2. \, V.G)]$

Component $K = G. \cos d$

$R_W = W . (\cos(y+1) - \cos y)$

$F_D = m . g . L_f / L_d$

$F_C = m . g . L_f . \sin x$

angle $b = \text{Tan}^{-1} (F_C / F_D)$

angle $e = b - d$

$F_R = F_D / \cos b$

$F_{GT} = F_R .\cos e$

$F_{GC} = F_R . \sin e$

$R_y = F_C .180 / (m . K . pi)$

$R_K = F_{GT} .\cos d / m$

$R_H = R_y . R_W$

$R_V = R_H + R_K$

Values of R_K, R_H and R_V are plotted versus the range of angles of bank. In this example a range of angles of bank are tested at one wind-angle in the windward turn producing the result in figure 16.4.

Albatross dynamic soaring

To test the Windward Turn Theory, the whole dynamic soaring manoeuvre must be plotted. To do this, the spreadsheet is written with a single angle of bank for each turn but over a range of wind-angles. To get the result in chapter 13, figure 13.10, the windward turn is given a fixed angle of bank of say 10 degrees and a suitable range of wind-angles say 120 through 90 to 060. The leeward turn is given a fixed angle of bank of say 60 degrees and the same range of wind-angles. The program has a logic which flips the bank angle from windward to leeward at the appropriate wind-angle. The starting data is airspeed and height. Wind is uniform.

Using **V**, **W** and **y**, groundspeed, drift and component **K** are calculated. **Ry** is calculated which gives the time interval between say 10 degree intervals of wind angle. **R**$_H$, **R**$_K$ and **R**$_V$ are calculated, then the new **V** which is carried forward to the next station. Angles of climb or descent are nominated to create the level windward turn and leeward wing-over. Height is calculated at each increment along with appropriate acceleration due to gravity. The end result is inspected and should give a net gain of airspeed and height.

The spread sheet can be designed to work in increments of wind-angle, time or distance, according to which of the dynamic soaring manoeuvres is being demonstrated.

This boils down to the fact that we can use the same set of equations to describe the bird motion in both the windward and leeward turns. The only differences are that the angle of bank has a negative sign in the windward turn and a positive sign in the leeward turn and we have to nominate the load-factor in the leeward wing-over.

Two mathematical solutions

Using the components **K** and **H** is a good way of visualising what is going on; but is there a better way of calculating the acceleration of **V**?

Another way is to take the triangle equation and differentiate to get an equation for the rate of change of airspeed. This is more accurate but it is less easy to visualise what is going on. Looking at the triangle of velocities under acceleration in the windward turn, the wind-velocity is uniform but the air-velocity and ground-velocity are variable. (Figure 16.5). The unbalanced forces can act on both the magnitude and direction of the ground-velocity and then the rate of change of the ground-velocity affects the air-velocity. If the groundspeed decreases, the airspeed decreases and/or the wind-angle **y** decreases. If the direction of the ground-velocity (track-angle **z**) increases, then the airspeed increases and/or the wind-angle **y** reduces.

16.5 Triangle of velocities

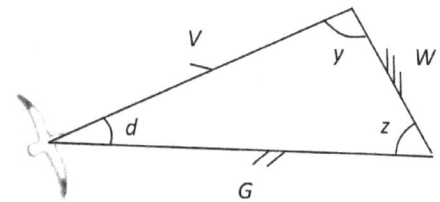

V   Air-velocity
G   Ground-velocity
W   Wind-velocity
d   Angle of drift
Y   Wind-angle
z   Track-angle

Using a derivative

Instead of calculating airspeed acceleration R_V by summing the incremental values of R_H and R_K, an alternative method is to take the derivative of the triangle equation with respect to time. This will give us an equation for R_V as a function of airspeed, groundspeed, track angle and wind-velocity, rate of change of groundspeed and rate of change of track-angle. The derivative gives the rate of change of airspeed at each wind-angle, which then gives the airspeed carried forward to the next increment.

Here is the derivative over time, dV/dt. The wind is uniform therefore the rate of change of the wind-speed and direction is zero, so these terms are deleted.

dW/dt = 0

$$V^2 = G^2 + W^2 - [2.G.W.\cos z]$$

$$2.V.(dV/dt) = (2G.dG/dt) + \cancel{(2.W.dW/dt)} - [d(2.G.W.\cos z)/dt]$$

$$2.V.(dV/dt) = (2.G.dG/dt) - [(2.G.W.(d \cos z/dt)) + \cancel{(2.G.\cos z.(dW/dt))} + (2.W.\cos z.(dG/dt))]$$

$$2.V.(dV/dt) = (2.G.dG/dt) - [(2.G.W.(-\sin z.dz/dt)) + ((2.W.\cos z.(dG/dt))]$$

$$V.(dV/dt) = (G.dG/dt) + (G.W.\sin z.(dz/dt)) - (W.\cos z.(dG/dt))$$

$$\underline{dV/dt = [(dG/dt.(G - W.\cos z)) + (dz/dt.(G.W.\sin z.))] / V}$$

162

This equation tells us that the rate of change of airspeed depends on the rate of change of groundspeed and the rate of change of angle **z**. (figure 16.5). Now we need force components that will give us these rates of change. See figure 16.2 and 16.3. Force component $\mathbf{F_{GT}}$ gives the rate of change **dG/dt** of ground-speed **G**. Force component $\mathbf{F_{GC}}$ gives the rate of change **dz/dt** of ground-track **z**. We are going to plot the results against intervals of wind-angle (say 10 degrees) during a full 360 turn. To determine the time interval, we can use the rate of turn $\mathbf{R_y}$ (same as **dy/dt**), which is produced by force $\mathbf{F_c}$ (normal to air-velocity **V**).

If **dy/dt** is positive, then **dz/dt** is negative and the sum of **dy/dt**, **dz/dt** and **dd/dt** is zero; because **y+z+d** is always 180. The rate of change of the drift angle **d**, **dd/dt** flips between positive and negative twice in 360 degrees of turn as the drift flips from right to left and back.

Results

The results for the circling powered aircraft are figures 16.6 and 16.7 The airspeed data is the same in both graphs but with different scales. These are similar to figures 12.7 and 12.8. While groundspeed varies by plus and minus the wind-speed (+/- 10), the variation of airspeed is much less (+ 0.5/-1.5), which is barely 2%. This is similar to the speed that will be lost in a turn due to the increased load factor compared to straight flight and so is easily overlooked in normal flight operations. We can infer that the effect of the rate of turn or the changing track-angle **z** is slightly greater than the effect of the changing groundspeed, although the effects here are entirely due to the *forces* acting on the aircraft.

16.6 Circling flight
Airspeed and groundspeed

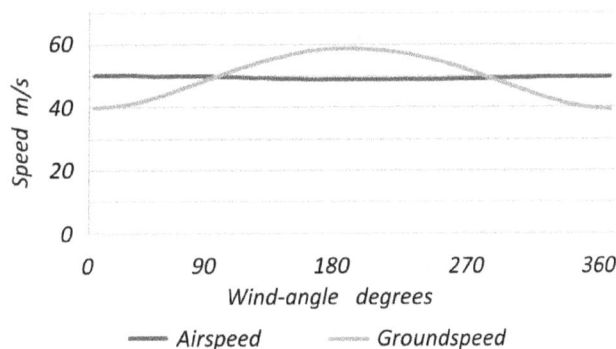

16.7 Circling flight
Airspeed

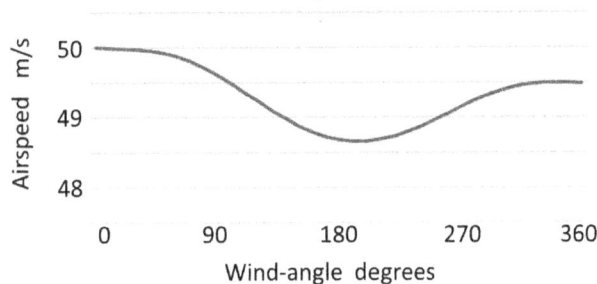

Alternative treatments

There are other ways to calculate a derivative. For example, the triangle equation can be written in three forms, one for each angle. The three equations can be rewritten for the airspeed term, then added together and the derivative taken. This produces a similar result to those shown here.

Exchanging speed and height

During the leeward turn the bird climbs and descends in a wing-over. An angle of climb will give an additional tangential force equal to the weight times the sine of the climb angle, affecting speed in the climb and descent. The height energy then converts to speed in the descent.

Vertical Acceleration

In the leeward wing-over, whenever the vertical component of lift is less than the weight, the bird will accelerate downwards. Its rate of descent will then be the sum of the effect of gravity and the effect of its angle of climb or descent. This limits the time and length of the leeward wing-over.

Height

In order to get a smooth transition between the windward and leeward turns the angle of climb and descent is controlled within nominated limits by a sine function. This gives a very shallow dive and climb in the windward turn and a more exaggerated wing-over profile in the leeward turn. Height is calculated using acceleration components determined by the angle of climb and descent and the vertical component of lift.

Bibliography

Alexander D E	Natures Flyers 2002
Brooke M	Far from Land 2018
Gill FB	Ornithology 2007
Henderson C L	Birds in Flight 2008
Jameson W	The Wandering Albatross 1958
Kermode AC	Mechanics of Flight 1972
Lilienthal Otto	Bird Flight… 1889
Onley D Scofield P	Albatrosses, Petrels... 2007
Pennycuick CJ	Gust Soaring 2002
Pennycuick CJ	Modelling the Flying Bird... 2008
Proctor NS Lynch PJ	Manual of Ornithology 1993
Rayleigh	Soaring of Birds 1883
Ruppell G	Bird Flight 1975
Sachs G	Minimum shear wind strength... 2005
Sachs G	In-Flight Measurement of Dynamic Soaring… 2010
Sachs G	Experimental Verification of Dynamic Soaring... 2013
Safina C	Eye of the Albatross 2002
Tickell WLN	Albatrosses 2000
Videler J J	Avian Flight 2010

Index

www.ingramcontent.com/pod-product-compliance
Lightning Source LLC
Chambersburg PA
CBHW051656210326
41518CB00026B/2606